Veins of the Desert

A Review on Qanat / Falaj / Karez

RIVER PUBLISHERS SERIES IN CHEMICAL, ENVIRONMENTAL, AND ENERGY ENGINEERING

Indexing: All books published in this series are submitted to the Web of Science Book Citation Index (BkCI), to SCOPUS, to CrossRef and to Google Scholar for evaluation and indexing.

The "River Publishers Series in Chemical, Environmental, and Energy Engineering" is a series of comprehensive academic and professional books which focus on Environmental and Energy Engineering subjects. The series focuses on topics ranging from theory to policy and technology to applications.

Books published in the series include research monographs, edited volumes, handbooks and textbooks. The books provide professionals, researchers, educators, and advanced students in the field with an invaluable insight into the latest research and developments.

Topics covered in the series include, but are by no means restricted to the following:

- Energy and Energy Policy
- Chemical Engineering
- Water Management
- Sustainable Development
- Climate Change Mitigation
- Environmental Engineering
- Environmental System Monitoring and Analysis
- Sustainability: Greening the World Economy

For a list of other books in this series, visit www.riverpublishers.com

Veins of the Desert

A Review on Qanat / Falaj / Karez

Ali Asghar Semsar Yazdi

Majid Labbaf Khaneiki

UNESCO-ICQHS

River Publishers

Published, sold and distributed by:
River Publishers
Alsbjergvej 10
9260 Gistrup
Denmark

River Publishers
Lange Geer 44
2611 PW Delft
The Netherlands

Tel.: +45369953197
www.riverpublishers.com

ISBN: 978-87-7022-084-2 (Hardback)
978-87-7022-083-5 (Ebook)

To those who sacrificed their lives
in the dark underground tunnels
to quench the thirst of desert

Contents

Preface

When we fly over some arid countries, many craters in intersecting rows may catch our attention. What are they? This is a question that flashes through the visitors' mind, coming to some desert cities. They are not caused by volcanoes or meteorites, but built by tens of generations of people who had to seek a way to quench their thirst in such scorching deserts. If we accept that the main function of technology is to enhance the human ability to better adapt to environmental conditions, these craters are ranked among the technologies, which could make a big difference to the course of history. These craters are the excavated soil dumped around the shaft wells of qanats, which stretch for tens of kilometers. A qanat is a gently sloping subterranean canal, which taps a water-bearing zone at a higher elevation than the cultivated lands. Qanats have played a vital role in underground water extraction since ancient times. Qanats run across the desert like the body's veins, bringing life and prosperity to the people who used to live off the water flowing down. This technology is the focal point of the genesis of civilization in some parts of the world. The harsh environment drove people to invent the technology of the qanat and the know-how revolving around it. The qanat carries a tradition of science and technology, which used to be practiced in order to overcome the technical obstacles of its construction. Thus, the qanat is not only a means of irrigation, but it should be seen as a technical and cultural legacy, which deserves more attention.

Due to their distinctive features, Qanats discharge aquifer water continuously, so that users (farmers, settlers, nomads, etc.) can perfectly adapt themselves to water fluctuations affected by droughts and wet years. In other words, Qanats have always enjoyed compatibility with nature as a practicable means for the rational abstraction of groundwater, proving that our forefathers guaranteed the sustainable balance of water resources through their wise policies. Unfortunately, we disturbed this wise practice through excess mining of water using modern technology such as deep wells and electrical pumps, which are a threat to groundwater resources in arid and semi-arid zones, and now we can clearly observe a fast decline in water tables throughout those regions. Therefore, qanats as the main means of sustainable utilization of these aquifers should be taken into consideration. The question that may be raised is why the process of well drilling and groundwater depletion still continues, in the face of government support for the system of qanats. But the answer is completely clear: governmental measures are necessary but not sufficient. In other words, additional actions should be taken in a way that is complementary to the measures the government takes in favor of qanats. Here we realize how important integrated water resources management and the enhancing of public awareness can be. In fact, given the potential role that they can play in water resources management, all levels of society should be educated on the situation of water resources. This educational campaign should begin from the lowest level, or primary school, to the highest one, at the administrative scale. The water consumers, who are mostly the farmers, should receive a continuous and purposeful training program in order to stay informed about the situation of water resources.

It is time for society to believe in the great potential our traditional know-how has, which can be incorporated into our new irrigation systems. It is not wise to give up all modern technologies and revive and use only traditional methods, but it is quite wise to adopt the sustainable relationship, which has always existed between the environment and elements of the traditional production system. Therefore, there are many things we can learn from traditional irrigation to promote our modern water affairs. Recently, the attention of some NGOs, Green Groups, and non-profit institutions, along with governmental organizations, to traditional irrigation is on the increase. This gives glad tidings that future is not that bleak if we learn how to have both tradition and modernity living side by side meeting a unique purpose, and it is the golden key to the sustainable exploitation of groundwater.

This book takes up the issue of qanats in general and some other subjects revolving around this technology. In fact, what makes this book distinct from the other works on qanats is its emphasis on interdisciplinary studies, viewing all qanat-related issues as a whole. In this book we tried not to rehash what is already available in many references about qanats, and not to linger over such general facts as its definition or mechanism, but have described some new approaches, for example the impact of new strategies or technologies on the destiny of qanats, the measures some countries are taking to preserve qanats, how to better identify this environment-friendly technique and how to learn from the know-how accumulated about qanats over time, in order to devise a sustainable groundwater abstraction system.

Nevertheless, we confess that this book does not encompass all that should be said about qanats. For example, the role of qanats in political history deserves more attention, given that qanats underlie a civilization with all of its cultural, economic and political characteristics. The system bestowed a kind of economic and political style on the people living in this territory, which can be traced in political processes even today. This issue should be examined in a separate work in detail, and we hope we can redress this deficiency in near future.

Also, part of the book examines the situation of qanats, in different countries. This does not mean that those countries, which are not mentioned necessarily lack the qanat system, and the information on the countries mentioned in the book is limited to our present knowledge. Therefore, we welcome the reader's comments on the situation of qanats in their own country to complement this part in future.

Ali Asghar Semsar Yazdi

Majid Labbaf Khaneiki

Acknowledgements

We would like to acknowledge the local qanat masters who generously made time to participate in our interviews. Also, special thanks go to Ms. Jacky Sutton the UNESCO expert who edited the last chapter "Interviews with Local Practitioners" with patience and accuracy just for the sake of qanats and their promotion, and we thank Ms. Maryam Hatam-poor for her nice cooperation. We also thank Mr. Mustafa Shafiyi Kadekani for his paintings that well illustrate Chapter 6. For Chapter 6, we are indebted to the efforts our colleagues made and their helpful cooperation, we sincerely thank them all; Mr. Abdolazim Pooya, Mr. Mohammad Ali Amir Beyki, Mr. Morteza Tafti, Mr. Mohammad Hossein Alamdar, and Ms. Farhaneh Mehr Avaran.

Also, we express our heartfelt gratitude to Dr. Harriet Nash, British hydrogeologist and researcher, who patiently read through the book from cover to cover, and made a lot of valid and valuable comments, which brought this book to its final shape.

Special thanks go to International Center on Qanats and Historic Hydraulic Structures (UNESCO-ICQHS), Iran Water Resources Management Company and Tamadon Karizi Consulting Engineers Company, which facilitated this study from its very beginning.

Eventually we heartily thank Mr. Saleh Semsar Yazdi and Dr. Alireza Bazargan who kindly put us in contact with the River Publishes, paving the way for this book to appear at an international level.

List of Figures

List of Tables

1

Technical Overview

1.1 Qanat Definition and Components

The qanat is an underground gallery that conveys water from an aquifer or a water source to less elevated fields. In practice, a qanat consists of a series of vertical shafts in sloping ground, interconnected at the bottom by a tunnel with a gradient flatter than that of the ground. The first shaft (mother well) is sunk, usually into an alluvial fan, to a level below the groundwater table. Shafts are sunk at intervals of 20 to 200 meters in a line between the groundwater recharge zone and the irrigated land. From the air, a qanat system looks like a line of anthills leading from the foothills across the desert to the greenery of an irrigated settlement. The different parts of a typical qanat are shown in the Figure 1.1.

The parts of the qanat are described as follows:

Gallery (Channel): the qanat gallery, as shown in Figure 1.2, is called in Persian “rahrow” or “kooreh”, which is an almost horizontal tunnel dug to get access to groundwater reserves, and to transfer this water to the earth’s surface. The dimensions of the gallery are such that the workers can easily go through and work in it: between 90 and 150 centimeters high and its width is less than half the height.

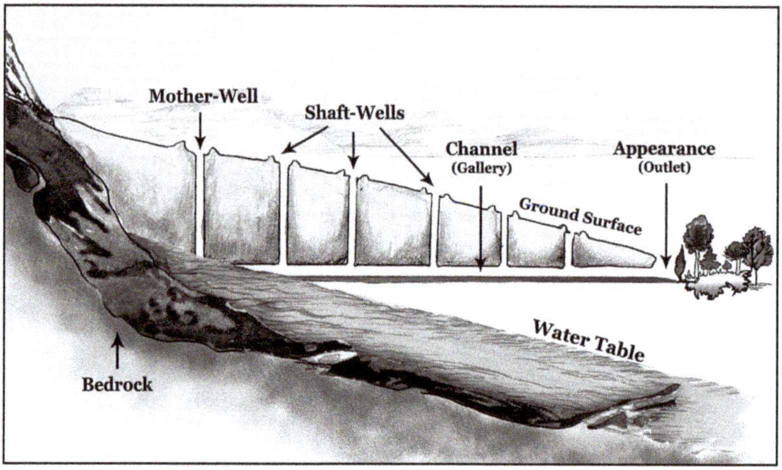

Figure 1.1 Longitudinal section of a qanat.

Figure 1.2 Gallery of a Qanat.

Taking into account the topography and position of the water table, the tunnel gradient is calculated so that the water flowing in the tunnel would have an adequate speed not to leave too much sediment and, on the other hand, would not be so rapid that it washes away the tunnel and causes erosion. The optimum gradient is taken to be between 2/1000 and 5/1000. The length of the tunnel depends on a variety of factors such as the topography, aquifer behavior, qanat discharge, where the water is to be used, etc. A qanat may be

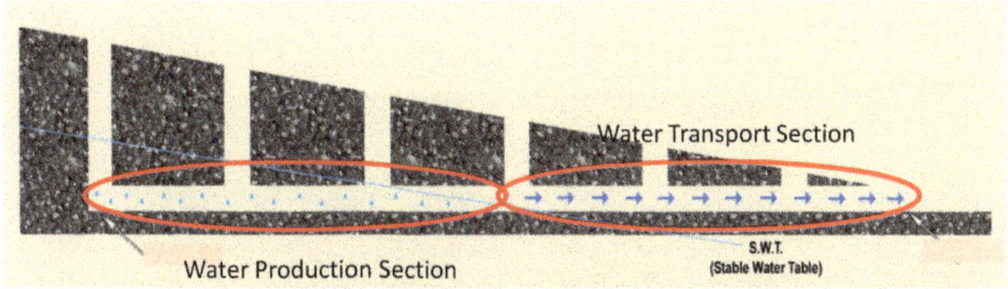

Figure 1.3 Water production and transport sections.

a few hundred meters to about one hundred kilometers long. The greater the extent of qanat in the water-bearing zone, the more water infiltrates into the tunnel. Hence, a qanat may have several side branches in the water-bearing zone to increase the discharge.

Water production section: That part of the qanat tunnel, which cuts through the aquifer and collects the groundwater seepage is called "tare-Kar" or water production section (Figure 1.3). This part is below the water table and is fed by the aquifer. The discharge of the qanat completely depends on the extent, permeability and the depth of this part in comparison to water table. The length of the "tare-Kar" is not fixed but correlates with the fluctuations in water table.

Water transport section: That part of the tunnel which simply conveys the collected water to the surface is called the "khoshke-kar" or water transport section. In this part of the qanat; some water is lost through percolation; hence, the more impermeable this part, the less water percolates back into the aquifer. There are some methods to waterproof this section, one of which is to line the tunnel with concrete or ceramic hoops called "kaval". It is worth noting that another application of such hoops is to shore up the tunnels to prevent any possible cave in.

Transition zone: The border between the water production and transport sections is called "hadd-e taran va khoshkan" or transition zone.

Outlet (appearance or exit point): Where the tunnel and the ground surface eventually intersect is the outlet of the qanat, called the "mazhar", implying where water appears (Figure 1.4).

Shaft wells: There are some vertical shafts sunk along the tunnel to connect the surface to the horizontal gallery. The main application of these wells is to haul the debris and excavated materials from the tunnel on to the surface

Figure 1.4 Qanat exit point.

while the qanat is under construction. They also provide access and help ventilate the tunnel and provide more oxygen for the workers when being repaired. These wells play an important role in repairing the qanat, by making it possible to send down the needed facilities and tools and remove the debris. The shaft wells cut short the time needed for qanat construction or repair and reduce the relevant expenses. A shaft well is between 80 and 100 centimeters in diameter, and the distance between the wells vary from 20 to 200 meters. In fact, the deeper the shaft wells, the further they are from each other. The distance between the shallow wells is often two times as their depth. If a shaft well is sunk in a soft crumbling soil, to prevent any collapse, the well would be shored up with brick or stone linings or concrete or ceramic hoops.

Mother well: The farthest shaft well from the outlet sunk upstream is called the "mother well". The mother well is usually the deepest well, in which a large inflow of water shows that the qanat is in a satisfactory state. If the water table goes down so much that it is located below the bottom of the mother well, no water can seep into the gallery, and if this situation persists, the qanat will inevitably dry up. If a qanat is extended so far that another well is needed, the new well would now be the mother well and the former one would be a normal shaft well. In a nutshell, the last well is always called the mother well. The depth of the mother well varies from qanat to qanat,

Figure 1.5 Farmlands irrigated by qanat in Bam.

and the deepest one in Iran has been estimated in the qanat of Gonabad to be 300 meters.

Irrigated land: The farm is a cultivated area which is less elevated than the outlet of the qanat, irrigated by the water coming out of the qanat (Figure 1.5). The extent of the cultivated area depends on several factors such as the qanat discharge, soil quality, soil permeability, local climatic conditions, etc. If the water flowing from the qanat is insufficient, the water is stored in a pool to increase the volume and the head of water so that it can be delivered to the land at a higher flow rate and thus irrigate the farms.

The irrigation cycle differs from area to area but is usually between 12 and 15 days. It should be noted that an irrigation cycle is a time-based water management system according to which the shareholders take turns irrigating their farms. For example, if the irrigation cycle is 12 days, every farmer has the right to take his share just once every 12 days.

The mechanism and morphology of qanats vary from region to region with geographical and geological conditions. In the following topic, we examine the different types of qanats we have come across so far.

1.2 Qanat Classification

We can classify qanats according to the following five criteria: length and depth of qanat, topography, geographical situation, qanat discharge, and source of qanat flow, as shown in Figure 1.6.

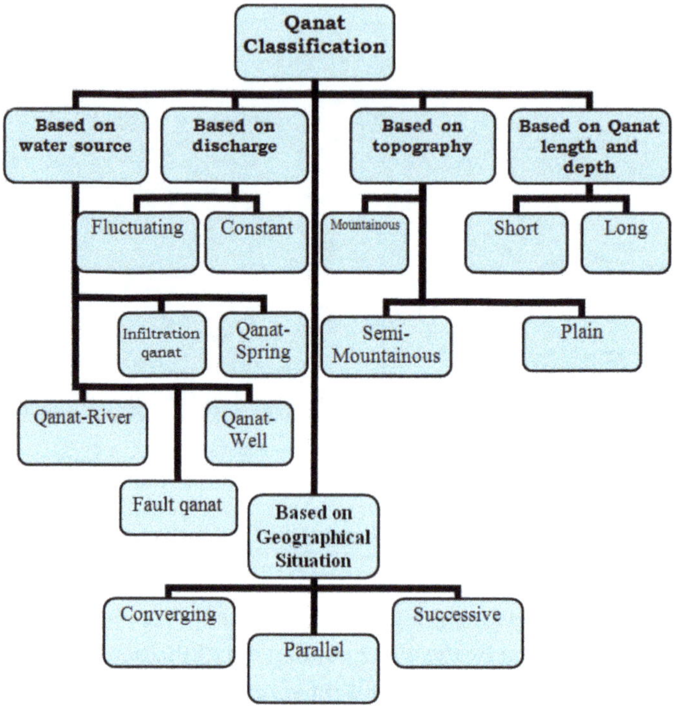

Figure 1.6 Classification of qanats.

1.2.1 Classification according to Length and Depth of Qanat

This criterion is the most common one to classify qanats and in fact correlates with the region's topography. According to this criterion all the qanats can be classified into two categories; short and long. The short qanats usually run along steep valleys or down the mountains foot, and their galleries intersect the surface after a short distance, while long qanats run across the plains, and so their galleries have to go further to reach the surface. Long qanats have deeper mother well.

1.2.2 Classification according to Topography

Taking into account the contour lines on a topographic map, qanats can be classified into the three groups of mountainous semi-mountainous and plain qanats, each of which has special traits.

A. *Mountainous qanats:* One of the general characteristics of such qanats is having a shallow mother well and consequently shallow shaft wells, because in these locations the impermeable layer is not so deep, thus the water-bearing zone is not far from the surface. The shallowness of the shaft wells and the steep slope of the ground surface lead to the gallery of such qanats being relatively short. The mountainous qanats are called “hava-Bin” by the Iranian qanat practitioners, which literally means “looking to the sky”, implying that their discharge fully correlates with the amount of rainfall: their discharge is high during a wet year and vice versa, although we may find some mountainous qanats with a more constant discharge because they draw on more extensive groundwater reserves in faults or fissured rock (Figure 1.7).

B. *Semi-mountainous qanats:* Semi-mountainous qanats have a mother well that is sunk in a mountainous area but have an outlet in a plain. These qanats are longer and enjoy deeper mother wells than the last type, and so their discharge is higher and more reliable (Figure 1.8).

Figure 1.7 Mountainous qanats.

Source: (Semsar and Labbaf, 2017).

Figure 1.8 Semi-mountainous qanats.

Source: (Semsar and Labbaf, 2017).

C. *Plain qanats:* These qanats are longer and have deeper mother wells than the last types. Given that in the plain area, the impermeable layer is deeper and the surface slope is gentler, the groundwater reserve is present at a lower depth. Hence, to gain access to the aquifer, the qanat workers have to dig the tunnel far deeper, leading to a longer qanat. These qanats have a higher and more reliable discharge in comparison to the previous types.

1.2.3 Classification according to Geographical Situation

This type of classification reflects the location of qanats in comparison to each other, and closely correlates with the topography. Based on the geographical situation, there are three types of qanats, as follows:

A. *Successive qanats:* Successive qanats are usually seen in mountainous areas and along steep valleys, and the layout is such that an upper qanat can replenish a lower one. In other words, the mother well of all but the highest qanat is sunk in the land irrigated by an upslope qanat, tapping the percolation of irrigation water. If this type of qanat is present in arid regions, the lower qanats will discharge less and less water, because the lower qanats just drain out the water escaping from the upper qanats. By contrast in semi-arid and humid regions the downstream qanats not only tap the groundwater but also collect the percolation of water coming from the upper qanats.

B. *Parallel qanats:* This type of qanat is usually found along the foot of mountains, stretching out from a mountainous area to a plain nearby. These qanats may run along an eroded valley whose bed is filled up with alluvial materials. They are fed by the rainfall seeping into the surrounding mountains. Hence, the water seepage into their galleries is more in the upstream at the end of the valley than in the downstream.

C. *Converging qanats:* These qanats are seen in the plains encircled by the mountains. These concentric qanats follow the direction of groundwater flow from the surroundings to the center.

1.2.4 Classification according to Qanat Discharge

Another criterion by which qanats can be classified is their discharge. Qanats may enjoy constant or variable discharge, depending on the type, and extent of the aquifer. The mountainous qanats, which are fed by shallow aquifers enjoy fluctuating discharge. Thus, such qanats discharge more water after a rainfall and less water when precipitation decreases. In contrast, the plain qanats, which cut through deep aquifers are less affected by such fluctuations in rainfall. We can also classify qanats according to whether they are fed by a confined, semi-confined or free aquifer.

1.2.5 Classification according to the Source of Qanat Flow

This criterion to classify qanats may seem strange, because it is taken for granted that the flow of every qanat originates from groundwater reserves directly. However, the reality is that in some regions the water discharge of the qanat may not be the groundwater seepage into that qanat's gallery, but at least some of it may come from a nearby spring or river. Hence, this criterion examines whether the water is the direct result of the groundwater infiltrating into the qanat tunnel or is coming from other sources like rivers. Thus, we can have five types of qanats as follows:

A. *Infiltration qanat:* An infiltration qanat drains groundwater, which directly enters the qanat production section from a homogenous aquifer. Most of the qanats fall into this category.
B. *Qanat-spring:* When water from a spring enters a qanat and adds to its water, such qanats are called qanat-springs. A good example is the qanat of Ahrestan running in Taft near Yazd, which receives the water of the spring of Tamer and delivers all water at a point downstream. It is worth noting that there are also some qanats that directly hit a fault and drain water from the fault zone. Such qanats can also be categorized as a "qanat-spring".
C. *Qanat-river:* This qanat resembles the last type, except that a qanat-river receives a surface stream (whether permanent or temporary; Figure 1.9). In most cases, it is impossible to transfer river water to the desired lands by gravity through open trenches, because topographical conditions may rule out the possibility to transfer the river water simply by an open canal. Therefore, digging a subterranean conduit is helpful to sort out the topographical problem. The structure of this conduit resembles that of a typical qanat, although its water has nothing to do with an aquifer. There are many qanats of this kind in the province of Khuzestan in Iran, which are locally called "qomesh".
D. *Qanat-well:* Qanat-well is a combination of the technique of a pumped well and a qanat. When the aquifer drawdown is so much that the horizontal extension of the tunnel does not improve the flow, or the

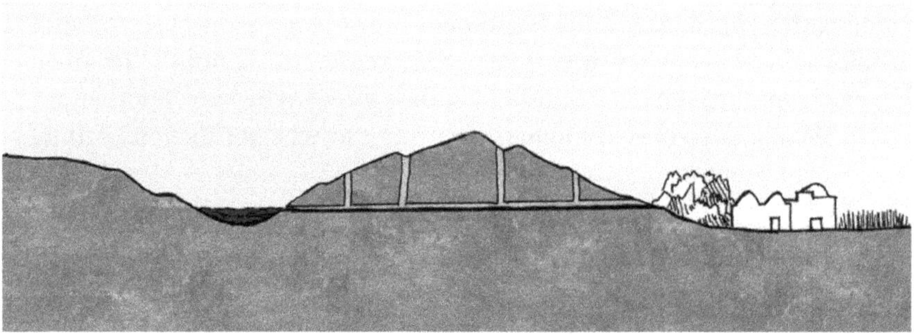

Figure 1.9 Qanat – river.
Source: (Semsar and Labbaf, 2017).

Figure 1.10 Fault qanat.
Source: (Semsar and Labbaf, 2017).

geology is unsuitable for an extension, the only option may be to deepen the mother well to get access to the water-bearing zone. In this case, the deepened mother well is equipped with a pump to deliver the water up to the gallery.

E. *Fault qanat:* Fault qanats are referred to as qanats that cut across a fault where groundwater flows (Figure 1.10). In this case, the water changes its course from the fault to the qanat whose main water source comes from the same fault.

1.3 Building and Repairing of Qanats; Tools and Methods

To construct a qanat, first, an area, usually near a mountain slope, is chosen in order to dig the first well. The qanat practitioners continue to dig the first shaft well until they come across the aquifer seeping into the bottom of the well. Then they stop digging the well, because of the rising water level and start digging a long tunnel crossing the bottoms of the wells, from the surface usually near the area to be cultivated, to the last and deepest well. The tunnel is roughly horizontal, with a gentle slope to allow the water to drain out.

The traditional qanat practitioners have a rich resource of knowledge passed down from generation to generation. When they want to locate a water-bearing zone in a desert to construct a qanat, they focus on some indications leading to the best spot. According to them, there are many signs, for example, the type of plants and soil or a river bed, which indicate the presence of groundwater. Broad alluvial valleys with thick sedimentary layers and underlying solid bedrock are the best sites for qanats. They analyze the soil, as this affects the quality of the groundwater, and then sink a trial well in the middle of the valley where the two alluvial slopes meet. Then they wait until the groundwater seeps into it and monitor the seepage rate to make sure that the permeability of the surrounding layers is adequate to support a qanat. If it is, they start to dig the next well, at a lower elevation and the horizontal tunnel between the wells. If the permeability is not adequate they abandon that well and sink another one elsewhere.

The qanat practitioners use some methods to determine the length of a qanat and locate its outlet. First, they calculate the difference between the levels of water in two different wells, and then they use a leveling tool to make sure that the spot where the trial well is to be dug is higher than the point where the gallery is to exit. This is how they know the exact location of the exit of the qanat. In Chapter 6 "Interviews with Local Practitioners", the leveling tool has been described in detail.

In most cases, the distance between the shallow wells is often two times as their depth; for example, if a well is 100 meters deep, it would be 200 meters away from the next well. The maximum distance between the wells can be twice their depth. If the wells are far apart, they have to dig the twin wells in order to ventilate the tunnel. Twin wells facilitate circulation of fresh air in the tunnel and consist of two wells just 5 or 6 meters apart, which are called "jofte badoo" wells in Yazd, Iran. They are connected by some short tunnels that help the air better circulate. Each well is at least 78 cm in diameter to allow room for the well digger. He will use a short-handled pick for this work and will climb up and down the well using niches dug into the well side, forming a kind of ladder, known as a "paraf".

A question that may come to mind is how they dig the tunnel so straight and true that it intersects the vertical shafts at their bottom. To do so, the workers use a plumb line made of a wooden stick with two stones on ropes tied to each end. They place the stick over the mouth of the well, such that at the bottom, the stones would be orientated toward the right direction. The person in the tunnel holds a light behind one of the stones so that there is only one shadow on the opposite wall, and this points the workers in the right direction.

If a water-saturated layer of soil lies above a tunnel, it is obvious that the workers cannot dig a well through it, because the water would seep into the well and the build-up of water easily hinders digging. In this case, to get the shafts to the tunnel, the workers resort to a very amazing but dangerous method called "devil-kani", which involves digging the well from the tunnel to the surface such that water pours into the tunnel and drains out. This means that work can continue, and in fact devil-kani is a technical feat in qanat construction.

Apart from this procedure, the workers may face many other dangers such as poisonous gases. If the gas is not so much, they hold a piece of canvas on to their mouth and nose. Otherwise, they dig a twin well, or use a kind of bellows made of leather by means of which they can send the fresh air down the well. There is a hose linked to the bellows that conveys the air to the bottom of the well. Another option is to drop sand, vinegar or lime into the well. When sand is dumped into the well, it pushes up the air at the bottom of the well and fresh air replaces the air coming up. Another danger to the workers is collapse of the shafts or tunnel/gallery. To prevent the shafts from collapsing, they line the shaft from the bottom to the top with stone, brick, and ceramic hoops, although nowadays, they also use cement and install metal rings inside and along the shaft to line it. They also drive a round metal mold into the well, and then fill the space between the mold and the well with reinforced concrete and iron rods. In case the gallery roof is weak and even caves in, they install concrete scaffolding rings to support the tunnel or, if the risk of collapse is not too great, they line the sides and ceiling of the tunnel with pieces of rock. Also, there are more details about what facilities they have to insure the welfare and safety of workers. For example, there is a tiny chamber built into the side of the shaft about two meters below the surface where the qanat practitioners can rest or change their clothes. There is a similar space at the bottom of deep shafts to provide shelter from debris falling from the bucket as it is being pulled up. In this regard, they also hold a round wooden plate above their heads like an umbrella to protect themselves.

One of the disadvantages the critics attribute to the qanat is that this system continuously drains groundwater, even when there is no need for water as is sometimes the case in winter. But the qanat practitioners traditionally

Figure 1.11 A typical underground dam from two angles (front and side).

know how to prevent winter water from being wasted. Where the qanat cuts through hard soil, it is very easy to build an underground dam (Figure 1.11). This is a brick wall built across the tunnel where the water production section ends, and the water transport section begins. There are four or five outlets sealed with bricks and when the farmers need water, they just pull the bricks out.

Also, to prevent damaging the aquifer, the qanat masters delineate an area surrounding the qanat between 1 and 3 kilometers in diameter, depending on the local conditions, as protected.

A qanat is an extended system, which is subject to a variety of detrimental factors, from crumbling soils to water pollution, so a qanat system has to be repaired and cleaned once in a while. The most important factor in the repairing and cleaning of qanats is manpower. A qanat construction team is made up of between three and six people, depending on the length of the qanat and the depth of the wells. There are four distinct categories of qanat practitioner: the "karshenas" who decides where the new branches shall be dug and supervises the entire project; the "ostad kar" or master worker, who is in charge of digging; the "gelband" who collects the debris, and the "charkh kesh" who operates the pulley to haul the bucket up the well. If the well shafts are far apart, another person, the "lashe kesh" is also added to the team to drag the bucket of debris along the tunnel to the nearest shaft and to tie the bucket to the pulley rope.

The qanat practitioners use a pickaxe, pulley, rope, shovel, oil or carbide lamp, electric fan, compass and plumb line to construct the qanat, and they use a windlass to haul the debris from the tunnel. There are different types of pickaxe depending on the use: to dig at the end of the tunnel, they use a pick weighing 2–3 kg with a handle 50–60 cm long; for dredging the tunnel, they use a pick weighing up to 6 kg with a handle up to a meter long. The type of pickaxe also depends on the soil: for hard soils they use one which has a short, thick point, rather like the beak of a sparrow, and they use a pick with a long thin point in soft soils. The handles are about 40–50 cm long if they are to be used for sinking a shaft and 50–60 cm if they are to be used for digging a tunnel and 90–100 cm if they are to be used for removing obstructions from the bottom of a tunnel. The leveling tools are two wooden poles 1.5 m long with a square base and a hole in the middle with a plumb line in it. The steeper the gradient, the shorter the distance between the two poles.

In general, the diameter of the pulley is around one meter, but that depends on the depth of the shaft. The deeper the shaft, the greater the diameter of the pulley. They reinforce the legs of the pulley by digging holes on both sides of the mouth of shaft, putting the legs into the holes and filling them up with stone and earth. They use cotton, palm fiber, steel wire, and plastic for the rope; obviously the deeper the shaft, the thicker the rope and the stronger the material that it is made of. They use steel wire for the electric pulleys. In the past, they used a kind of bucket of tanned skin of sheep, but this has now been substituted with a rubber bucket.

In the past, the only way to light the tunnel was to use a lamp like a small bowl full of vegetable oil in which there was a wick. But now they use a carbide lamp or DC electricity.

To repair and rehabilitate a qanat, two measures are commonly taken in IranIran. The first is to extend the horizontal tunnel to increase the seepage area of the qanat, and the second is to clean the tunnel periodically. Extending the horizontal tunnel which is called in Iran "pishkar-kani" may be either in the same direction or in different directions forming side branches. A side branch (dastak) serves to increase the discharge of the qanat when necessary. Side branches vary in length from 10 to 200 meters and the longer ones also have vertical shafts. Most qanats have two side branches running in opposite directions, and no additional branches are dug while existing branches are in use. If a tunnel needs additional shafts, the qanat practitioners dig the shaft so straight and true that it intersects the horizontal tunnel at the bottom, as described earlier.

To find the right direction, they used to use a special tool called a "rassi". This is a wooden stick with two stones suspended from each end on two pieces of rope. The stick is placed over the mouth of the well pointing at the direction of the gallery, and the stones are allowed to hang down the well like a plumb line. The orientation of the two hanging stones down the well tells them where to dig forward in the tunnel. Nowadays they use a compass instead.

Amid the process of extending a qanat, the qanat practitioners may enter the vicinity of another qanat. No qanat should run under the water production section of another qanat, otherwise the lower one may/will drain the available groundwater from the upper one. To determine whether a qanat cuts under the water production section, they gouge a hole in the wall of the gallery to see if there is any seepage in it. If there is none they allow the lower qanat to keep making its way upslope. Though they are so tenacious in preventing qanats from draining each other's water, sometimes they come to an understanding to use a qanat gallery for transferring another qanat's water. If a qanat is severely damaged and is under repair, the owners of a neighboring qanat may agree to help the owners of the damaged qanat by allowing them to use the healthy qanat to transfer their water. To do so, they connect the two qanats with a side branch or a burrow called a "gozare", and the owners measure the volume of water entering the healthy qanat so that they can deliver the same amount of water downstream.

Sometimes, the groundwater level sinks so fast that the qanat practitioners have no opportunity to extend the tunnel, but they have to deepen it. If the drawdown of the water table continues, even deepening the tunnel no longer works, and they start to dig another tunnel under and parallel to the old one. They use the shafts of the old qanat to do this and the practice is known as "tahsoo-bonsoo".

The second job of maintenance, which is commonly done in the qanats is to clean the tunnel once a while. They dredge the tunnel usually once every one or two years, unless the qanat runs through hard rock, when it's just once every 10 or even 20 years. The higher the concentration of dissolved minerals in the water, the less often they need to dredge because the minerals form deposits that line the sides of the tunnel. They usually dredge the tunnel in the fall or winter, when the demand for water is low and the qanat's discharge dwindles.

Apart from the aforementioned measures, the qanat practitioners have other methods to preserve the qanats. For example, if a shaft has been sunk in the middle of a valley that is subject to seasonal flash floods, they must protect it. First, they line the well with brick, stone, and cement from the bottom to 12 meters below the surface, where a concrete lid is installed. From here to the surface, the well is filled with lime, clay, or brick to prevent surface runoff from leaking into the well and causing problems with erosion and brining debris into the system.

Also, the qanat practitioners have some initiatives to reduce percolation in the water transport section. To waterproof the tunnel, if the tunnel is shallow, they dig and remove the ceiling of the tunnel, line the floor with cement and build two walls along both sides. A new method being practiced nowadays is to use bentonite (25 kg per 100 m) and sometimes even plastic sheets, but these methods can only be used in tunnels that are not eroded. Sometimes an earthquake may cause cracks in the tunnel through which water can escape. When this happens, they fill the cracks with very soft sand or clay. In this case, they mix soft clay with water and knead this mixture before spreading it on the floor and trampling it to fill up and seal any tiny cracks. They call this practice "koor kardan-e cheshm-e zamin" – "blinding the earth's eyes". If there are many cracks they also use ceramic or concrete hoops. Most of the qanat masters avoid this laborious business unless they have no choice.

Maybe one of the reasons why some old qanats have been deserted is their galleries cut through gypsum_layers in which the water percolation is high. To enhance the efficiency of water conveyance in the qanat tunnel, as well as to prevent any change in the water quality, from early on some methods have been practiced as follows: installing ceramic or concrete hoops, lining the floor and the sides of tunnel with concrete, laying pipes in the water transport section, putting a concrete half-pipe in the tunnel, spreading plastic sheets in the tunnel up to the level of flow, lining the water transport section with soft clay mortar, covering the tunnel floor with cement mortar, and using geomembrane sheets in the tunnel (Figure 1.12). Installing ceramic or concrete hoops is one of the common methods to reduce the water seepage from the qanat tunnel. This method not only reduces the water escape from the tunnel, but also reinforces these tunnels that are prone to collapse. In sum this practice has the following advantages: it can reduce the water percolation from the tunnel, it reinforces the gallery thus reducing the likelihood of collapse, and it can prevent any decline in the water quality while flowing through a saline or gypsiferous formation.

Figure 1.12 Qanat tunnel lined with geomembrane sheets.

Source: Dehghanpoor, Rasool.

The water production section of some qanats cuts through a sandy crumbling sediment in which the tunnel is subject to many collapses from time to time. In such qanats the water production section caves in completely over time, and only a small amount of water can flow through the mass of debris into the tunnel. Sometimes, the collapse is so much that the tunnel is completely obstructed and the water flow ceases. In such a case, we can put a flexible pipe in the water production section and then cover the pipe all the way with a special filter. In this manner, the water can enter the pipe and flow down to the outlet of the qanat. The construction cost of the flexible pipe is relatively high, but it is a permanent solution and can reduce the maintenance cost of qanat in long term.

Nowadays, one of the common methods at a lower cost is to use an impermeable geomembrane in the tunnel. This method can be carried out with less time and more easily than many of the traditional methods. The use of geomembrane sheets on the floor and the sides of the tunnel can dramatically decrease the amount of water percolating outwards, increasing the efficiency of water conveyance by up to 95 percent. The height of the geomembrane cover on the tunnel sides depends on the water level in the tunnel. The main agent used in such geomembranes is high density polyethylene, which is manufactured by modern machines in 7-meter-wide sheets of different thickness, but usually around 2 mm. Some chemical materials are added to improve its resistance to sunshine or other physical and chemical factors.

Geomembrane sheets can easily be spread on the floor of a qanat to keep the water from seeping into the ground in the water transport section. Before laying the geomembrane, it is essential to prepare the floor by removing any unevenness and plant roots from the gallery to prevent any damage to the membrane. The sheets are nailed to the sides of the tunnel. The application of geomembrane in a crumbling and soft tunnel is not recommended, unless the tunnel is lined and shored up prior to the use of geomembrane. The installation of one square meter of geomembrane, along with the preparing the tunnel, costs between 7 and 8 US dollars.

Another measure that can bring great benefit to the qanat system is to increase recharge to the aquifer. To do this, they build an earth dam or dyke to collect the seasonal runoff, which can then recharge aquifers. In mountainous areas, they build the dam about 150–200 meters upstream from the deepest and last well of the qanat so that the water trapped behind the dam will seep into the ground and increase the discharge. The dam must be far enough away from the qanat so that it does not affect its structural integrity. The softer the soil, the greater the distance between the two structures. The Iranians use qanats as a sustainable technique to extract groundwater, which is replenished in winters by some special earthen recharge dams locally called "*goorab*". The word consists of '*goor*' which means grave, and '*ab*' which means water. A goorab should be built upslope from the qanat's mother well and during the rainy season the runoff builds up behind the goorab and gradually seeps into the ground. Increasing the discharge of the qanat and controlling the surface erosion are two advantages of such dams. Sometimes, a goorab is more useful to a qanat than extending the tunnel. They also use abandoned qanats to recharge the aquifer by blocking the exit and directing the surface run-off into it through one of its shafts. The water seeps into the aquifer and replenishes the active qanats nearby.

1.4 Qanats in the Realm of Modernity

Just browsing through the contemporary history of some arid and semi-arid regions of the world can show the fact that while making a lot of progress in implementing modern technologies, these regions have incurred some losses. A backlash began against technical modernization, which manifested itself in environmental issues. For example, exploitation of groundwater with the aid of pumped wells placed pressure on the environment in arid and semi-arid areas. Before the advent of the pumped wells, the extraction of groundwater, mostly through the system of qanats, and natural recharge were in equilibrium. In fact, the technique of qanat construction which probably first came into existence in Persia about 2700 years ago evolved into a perfect state over time and turned into the main means to obtain groundwater in Iran as well as in many other countries. The advantage of the qanat was to drain groundwater without overexploiting the aquifer and leading to the depletion of groundwater levels. Hence, the outflow from and the inflow to the aquifer were usually in balance. Using this technique, people could take advantage of groundwater resources without disturbing its hydrological balance, but the introduction of the modern devices such as pumps and drilling machines changed this equation.

At the beginning, such modern devices received no welcome, but after drilling some pumped wells, the farmers increasingly turned toward these new technologies. When they compared the time it might take to construct a qanat (sometimes tens of years) with the time of drilling a well (less than one month), they admired this miracle of modern technology. When they wanted to increase the discharge of a qanat even a little, it took them two or three years to extend the tunnel, but it was easy to double the discharge of a pumped well just through changing the diameter of the pump or adding some stages to it. Hence, their tendency to use pumped wells increased greatly, with less attention to the environmental consequences of this technology. In this manner, most of the fertile plains were overrun by pumped wells, and vast areas turned into cultivated lands. But at this time, in some regions the growing number of pumped wells sounded the alarm about the depletion of groundwater which left many qanats dry. The water table kept descending, and the farmers had no way but to deepen their wells as long as the groundwater went down. Amid this problem, disputes broke out between the owners of qanats and wells, many of which went to trial. To put an end to this chaos, some water related laws were passed, which are still valid. In fact, over the past decades, pumped wells have played an important role in undermining the system of qanats. No one can deny that in some regions the only option to exploit groundwater is to drill wells, but this should not be used as a pretext to replace active qanats with pumped wells in other regions. Over-pumping can cause much long-term environmental and ecological damage of which the most important are:

1. The extraction of groundwater by means of a pumped well leads to the drawdown of the water table around the well, forming what is known as a cone of depression. Groundwater flows towards the well into the cone of depression. A well (due to the forced drainage) changes the natural direction of groundwater flow within the area of influence around

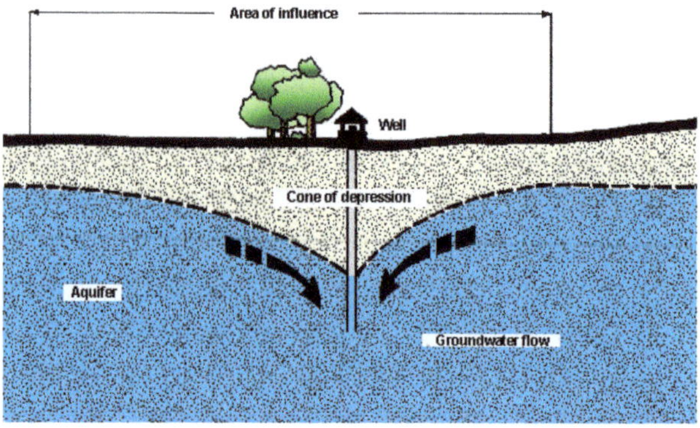

Figure 1.13 Drawdown of the water table around a well can cause a cone of depression.

Source: http://groundwater.oregonstate.edu/under/wells

the well. However, the qanat can discharge groundwater so gently that the cone of depression does not form or is almost imperceptible (Figure 1.13).

2. The drawdown of the water table caused by pumped wells, takes groundwater away from access by desert plant roots. This may end up in the extinction of some plant species and destroy the area's vegetation. Hence, the problem of desertification and wind erosion can intensify due to the lack of vegetation.
3. Pumped wells can drain the porous layers of an aquifer, resulting in the compaction of soil and even the occurrence of subsidence on the surface. In this case, some deep cracks may appear on the surface even affecting buildings in the area. In addition, the compaction of soil can diminish the capacity of the sedimentary layers to store groundwater, even though there would be later enough rainfall (Figure 1.14).
4. In several regions with Neogene formations, there is a layer of saline water below or surrounding the aquifer. The over-pumping through the wells sunk in such regions causes the saline water to creep into the fresh water zone, spoiling the groundwater quality (Figure 1.15).
5. In the regions where pumped wells cause depletion of groundwater or block groundwater flow, the surrounding qanats dry up.

1.5 Keeping Groundwater Drawdown at Bay

One of the most important problems in arid areas is the drawdown of groundwater, and controlling it is a very important measure to be taken. We have faced this problem in several deserts of the world; therefore, it is

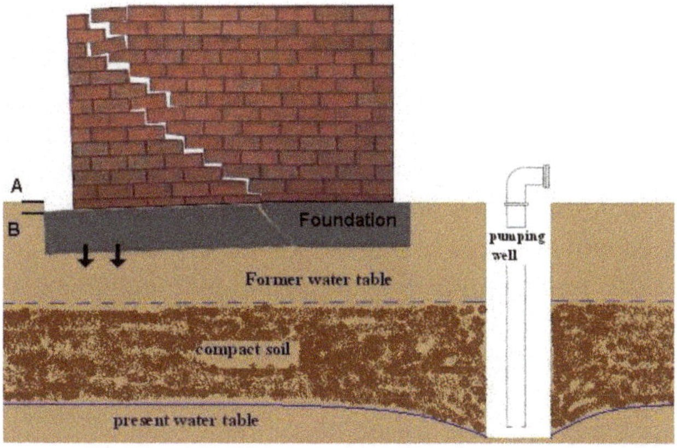

Figure 1.14 Overpumping of groundwater produces a drop in the foundation level from A to B and diminishes the volume of aquifer.

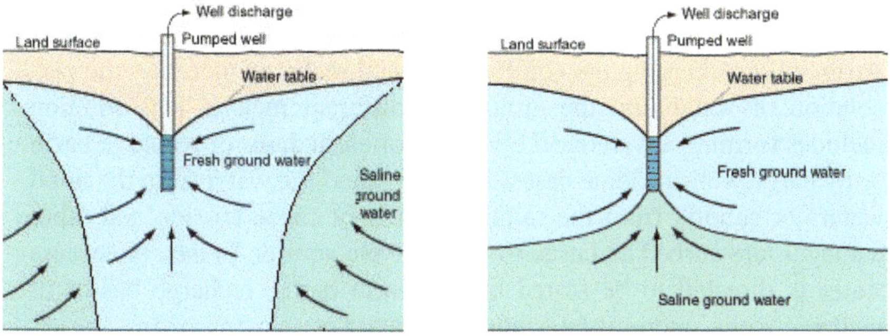

Figure 1.15 Large drawdowns in a pumped well can cause saline water to migrate into the well and increase the salinity of the water being discharged.

necessary to emphasize the role of water resources management. It is worth noting that although groundwater resources are decreasing, due to drought and overexploitation, the demand for water is still steady or even on the rise.

The main way to use water in the past was by means of qanats, and there was no groundwater drawdown. But as mentioned, after the advent of the new technologies for drilling and introduction of pumped wells, the natural recharge of the aquifer could no longer offset its discharge, so the water table went on declining over time, leading to poor quality water in most cases. Moreover, the drawdown of groundwater, and qanats drying up in the villages of arid areas in which people live off agriculture leads to an uninhabitable condition and to the people being evacuated from their villages.

The population growth over the past decades, decrease in precipitation in the wake of climate change and increase in the demand for water caused a boom in deep and semi-deep wells in arid and semi-arid areas of the world, and groundwater is being exploited even faster. However, the capacity of an aquifer is limited, and it needs enough time to be replenished through rainfall and infiltration. Therefore, water drawdown and the decrease in the capacity of aquifers have caused a critical condition in some dry lands.

To prevent more water drawdown, we should take important measures. In this matter, the four following solutions seem to be useful: Preventing irregular exploitation, artificially recharging the aquifer, preventing water outflow from the aquifer, and integrated water resources management. Here we examine these solutions, respectively:

1. Preventing irregular exploitation: In this case, firstly all the wells should have legal permission; secondly their total exploitation should be equal to the recharge of water in the aquifer. Subsequently, excavation of new wells or exploiting more water than what is allowed should be prevented.
2. Artificially recharging the aquifer: Artificial recharge increases the percolation of water into the ground by different means. This solution includes forming a watershed by means of earth dams or recharge basin or recharge wells in some cases. In this method the water from the flood waters or runoffs from the rainfall, that may cause erosion and other damages, are curbed and used to recharge the aquifer. In fact, the excess water is directed to be stored behind earth dams, recharge basins or recharge wells, and infiltrates into the ground.

    i. Earth dam: An earth recharge dam usually consists of several dams crossing the stream; behind the dam some of the water is stored and the rest overflows, reaching the next dam, which has the same function. By this method, the potential for infiltration reaches a peak.

    ii. Recharge basin: In this method, only one dam is constructed, which stores the water in a pool from which water percolates into the aquifer. But after a while, in both above-mentioned methods, some sediment settles behind the dam(s) and if it is not removed, water cannot percolate into the ground. This will also cause water loss due to evaporation. Therefore, the sediments behind the dams and in basins need to be removed so that they can work properly. Another way to avoid sedimentation is to construct sedimentation basins that are situated in the upstream of the main basin. In some

Figure 1.16 A typical artificial recharge dam in Fadagh, Fars province, Iran.

places, if the riverbed is suitable we can make the water whose sediments have settled flow to the riverbed and percolate into the earth, so this way the water table can be recharged (Figure 1.16).

iii. Recharge wells: A suitable place to construct dams upstream is needed for the two previous methods. For example, if the upstream lands are fertile, basins and dams should not be constructed, as they will remove valuable land from cultivation. In such cases it is better to use recharge wells. The advantages of this method are as follows: the water is directly injected into the water table and will not evaporate; the costs for equipment and excavation for this method are low; these wells do not need a large area. In this method the number of wells should be defined during geological and hydrological research and by taking into consideration, the depth of the aquifer, its permeability, etc. Recharge wells also silt up and therefore require maintenance from time to time.

Recently, a few artificial recharge projects using wells have been carried out in Iran. They have had considerable positive impacts on downstream water tables and the amount of discharge of the qanats.

iv. Underground recharge dam: This method means constructing an underground dam in the upstream of the aquifer, perpendicular to the groundwater flow. This dam reduces/stops groundwater flow and causes water to be stored underground. This prevents groundwater from being wasted and makes it possible to stored and direct the water in the desired direction and use it to recharge the aquifer. This kind of dam is often constructed to store groundwater in the rainy season and use it in the dry season. Needless to say, the building of such dams should be based on technical prudence: it is not possible to build such dams wherever we wish, but only where geological conditions are suitable. Needless to say, environmental issues should be taken into consideration in the construction of underground dams (Figures 1.17 and 1.18).

3. Preventing water outflow from qanat: This can be done by constructing underground dams inside the qanat gallery in a suitable spot to store water. Such dams are built to store water during winter when water is not in demand for irrigation.
4. Water consumption management: This concept in very crucial for integrated water resources management. Consumption patterns should be modified and regulated regarding agricultural, industrial and domestic sectors, in accordance with the available water resources and ecological capacities.

Figure 1.17 Sangoon underground dam in the village of Tamboo, Hormozgan province[1].

[1]This dam is 7.5-meters high from the bed rock and was built by the government with a budget of 150 thousand dollars.

Figure 1.18 Sangoon underground dam under construction, Hormozgan province.

5. Transfer of water between basins: This method is not recommended as a handy solution to solve water problems, though it is still deployed as a last resort. In Iran Inter-basin water transfer can be taken into consideration only to supply water for drinking and domestic sectors in critical areas; as a result, much migration can be prevented and also, it can help the aquifer not to reach the crisis point. As mentioned earlier, we should investigate and address environmental and ecological issues, which may come about in the wake of an inter-basin water transfer project.

1.6 Reactivating Dry Qanats

The arid and semi-arid zones of the world have had many qanats in the past, of which some are nowadays dry and inactive. To reactive the dry qanats, the first step is to understand the reasons why they dried up. These reasons can be classified into social and technical ones. The social aspects include effects caused by methods of qanat management, social conditions and new trends in human societies. The technical aspects include all the processes leading to drawdown of groundwater level or causing damage to the qanat's structure.

These include with a downturn in agriculture, farmers tending to change other activities; decrease in the number of technical experts in qanat related-affairs; and inefficiency in management and maintenance of qanats.

Here, we can mention an increase in the number of deep and semi deep wells; an increase in water demand; a change in cropping patterns; not maintaining or dredging qanats efficiently; considerable expense and long period needed for maintaining qanats; floods, qanat gallery's collapse, trespassing on a qanat's vicinity, waste of water in the gallery (outwards percolation), urban development, etc.

To cope with the mentioned social factors, it is necessary to show the importance of this historic hydraulic structure through some proper education and also to win the support of society for management, maintenance and reactivation of qanats through explaining the concept of sustainable development and comparing qanats with wells from this perspective.

To tackle the technical factors, it is essential to adopt a suitable solution for the main reasons for the qanat's drying up:

1. If the drying up or dwindling of qanat water is related to the drawdown of groundwater, the following technical solutions can be helpful:
   - Extending the gallery: By extending the gallery, the length of water production section increases and the longer part of the gallery cuts through aquifer, so that the qanat's discharge increases.
   - Displacing the mother well: This solution can be used if a rich aquifer is identified near a qanat. In this case, the direction of gallery is diverted towards the desired aquifer.
   - Deepening: Deepening the whole or a part of gallery's floor is called "kaf shekani". This is done to increase the qanat's discharge. In fact, by this action, the qanat water source is placed again inside the aquifer.
   - Artificial recharge: There are different ways to recharge qanats. These are classified into surface methods and underground methods. The surface methods include constructing obstacles on the ground surface to store the rain water on the ground, in these methods, water is stored on the ground surface and gradually penetrates the ground and reach groundwater.
   - Underground recharge means constructing an underground dam upstream from the qanat in the aquifer or inside the qanat. This dam reduces/stops groundwater flow and causes water to be stored underground, upstream of or in the upper party of the qanat. This prevents groundwater from being wasted and also makes it possible to direct the stored water in the required direction and use it to recharge dry qanats nearby. The dam can also be constructed under

the shafts or in the gallery of qanat. This kind of dam is often constructed to store groundwater in the rainy season and use it in the dry season. Needless to say, the building of such dams – whether underground or inside the qanat – should be based on technical prudence: it is not possible to build such dams wherever we want, but only where geological conditions are suitable.

2. If the drying up of a qanat is associated with the damage caused by floods or collapse of the gallery, the following solutions can be useful:
   - Dredging the gallery: removing mud and silt and also collapsed materials is called dredging. This work should be done periodically to prevent a reduction in the qanat's recharge. If not done regularly, the qanat's gallery may become blocked and water will not reach the exit. The qanat dries up and becomes inactive, little by little. In such cases dredging can be one of the most important solutions to reactivate the qanat.
   - Installing hoops: Installing hoops into the shafts and the gallery is called "kaval gozari". In some parts of a qanat, a cave in or collapse may happen, and may cause the gallery to become blocked and the qanat to dry up. So, it is essential to apply hoops on the walls of such qanats.
   - Lining qanat wells: This is done to protect the opening of shafts from collapse. It means to cover the walls of shaft from 3–4 meters below the surface to the top with some proper construction materials like stone, brick, etc.
   - Sealing up the well mouth: this should be done to prevent flood water from entering the qanat and destroying it. Flooding is perhaps the cause of the most severe of damage to qanats. The qanat practitioners used to explore along gullies in search of groundwater, and this method was usually helpful. In this case the qanat had to be built along the gully, which was subject to seasonal runoff, destructive to the qanat. Flood water entering the shaft well and flowing down the galleries can lead to destruction of galleries and shaft wells, and also after the settling of the water, sediments accumulate in the galleries and consequently the flow of water is obstructed (Figures 1.19 and 1.20). The build-up of water in the gallery may give rise to another collapse upstream. Moreover, another problem that may occur because of flood water entering the qanat is a decrease in the seepage rate into the qanat, due to

the soft residue left by the flood encrusting the seepage area. To keep flood water at bay, it is helpful to block the upper part of the shaft well, which can be done in several ways. Nevertheless, some wells should be left unblocked to have access to the qanat gallery; in this case a removable lid would be put on the well mouth then encircled with an earthen dike.

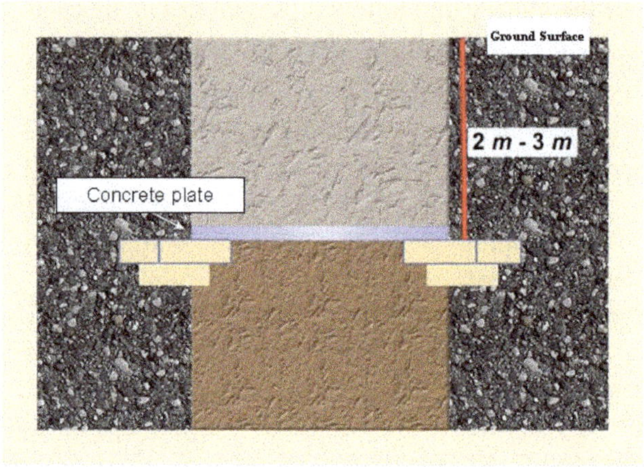

Figure 1.19 Blocking the upper part of the shaft well.

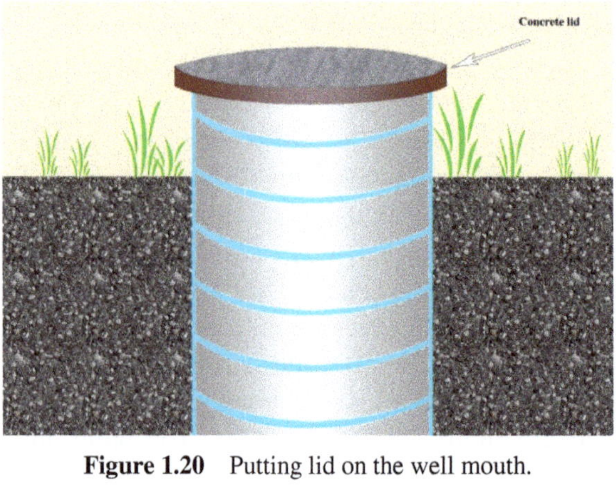

Figure 1.20 Putting lid on the well mouth.

1.7 Qanat-Dependent Structures

Qanat dependent structures are operational in close link to the qanat water, to fulfill other social needs than irrigation. Qanat related structures are ranked among the most interesting architectural structures, and most are relics of the past and are now considered part of historical heritage. The most important qanat related structures built in Iran are: watermills, "payabs", water reservoirs, pools, and "*bookans*".

1.8 Watermills

A watermill is a structure constructed to grind grain. Its main parts are a water house, two millstones, rotor blades and an axis, which connects the rotor blades and upper millstone vertically. The operation of the watermill is based on the potential energy of water due to the height of water house: The deeper the water house, the more energy is generated. Sometimes, the depth of a water house reaches 10 meters below the depth of the qanat to increase the water pressure (Figure 1.21).

In fact, water house is a shaft well that receives and accumulate water. When the qanat water reaches the water house, it pours down the shaft well and gushes out from a tiny nozzle at the bottom of the well and hits the blades, making the blades rotate, and imparting energy to the rotor, which then rotate

Figure 1.21 Watermill.

the upper millstone. The lower millstone is motionless. The friction between the upper and lower millstones grinds the wheat into flour. In some places, several watermills were operated by the water of only one qanat (Semsar and Labbaf, 2017).

1.9 Payab

A "*payab*" is a sloping gallery connecting the ground surface to the qanat gallery. This gallery has steps to make it possible for people to reach the water flowing in the qanat. The deeper the qanat the longer the payab. The slope of the payab is calculated so that the end of gallery meets the bottom of one of the qanat's shaft wells. The light was provided by the mentioned shaft. The size of a Payab stairway was such that two persons could go up and down side by side easily and their heads would not touch its ceiling. The payab was perpendicular to the direction of the qanat gallery to prevent the probable collapse of the gallery (Figure 1.22).

Inside the payab, the temperature is about 20° to 25°C, cool compared to ambient summer temperatures, due to its underground location and proximity to qanat water.

Some payabs were built for public use, near mosques, roads and caravanserais. But in some dry cities in central parts of Iran, many families had a private payab. The designers divided the qanat's main branch into several side branches inside the city. Each side branch crossed part of the city and then at the other side of the city, the side branches were joined to each other again. The houses neighboring the side branches had a private payab in which the owners rested during the hot days of summer. So, these houses were usually more valuable than other houses in the city. Also, there was a public payab in each part of the town for those who did not have access to a private payab, to use the water for sanitation. The payab structure was not complicated and the main part was like a room, which had a square or an octagonal plan with the following parts:

- There was a pool at the bottom of the payab. This round or polyhedral pool had some holes through which the qanat water entered or exited.
- Some platforms were constructed in the walls for people to sit on. There were more platforms in a public Payab than a private one. In the private Payabs, there were some shelves to put foodstuffs on. Also, there was a rope hanging from the ceiling above the pool, tied to a basket at the end to put some food, such as meat and fruit, to be kept fresh. The payab's roof was arched, resistant to collapse (Semsar and Labbaf, 2017).

Figure 1.22 Payab.

1.10 Water Reservoirs

The qanat water reservoir is an underground structure, constructed to store freshwater for domestic use. The reservoir was fed from a nearby shallow qanat. All the water reservoirs had a storage tank whose dimensions depended on the amount of qanat discharge and the demand for water. Most of the storage tanks were made of "*sarooj*"[2]. The different parts of a water reservoir were the storage tank, the roof of the storage tank, wind tower, stairway and ornamental portal.

The storage tank is of variable dimensions and its plan can be square, octagonal or a circle. The whole body of the water reservoir, or the most parts of it is constructed under the ground (Figure 1.23).

There were one or more faucets in the reservoir wall at 0.5–1.5 meters from the bottom. These faucets were used to transfer water from the storage tank to the "*pa shir*"[3] area, where people took the water through these faucets.

[2]Sarooj is a combination of lime, clay, and chipped straw, for cementing bricks or stones together.

[3]Pa shir is an area to tap water.

Figure 1.23 Water reservoir in Yazd, Iran.

The reasons why the faucets were set above the base are: to fill people's containers with water easily; to prevent the impurities deposited at the bottom of the storage tank, getting out of the faucets; and to have a high discharge because of the hydrostatic head in the tank.

The water storage tank's roof can be flat, but most of them are dome-shaped or conic. The conic roofs are two types; flat or step wise. The wind tower or "badgir" is one of the traditional structures, located in the water reservoir compound. Badgirs cause the air to circulate in the water storage area, evaporating the water and making it cool. The number of badgirs in a water reservoir ranges from 1 to 6.

One had to go down a stairway to reach a landing named "pa shir" to tap the water. The number of steps depended on water reservoir's depth. The pa shir was constructed at the same level as the bottom of the water storage tank, or a little lower. There was a drain at the bottom of the pashir to discharge the waste water to a canal named "rah ab" and this water was then directed to a nearby qanat. Sometimes, the water in the reservoir is polluted and can no longer be used, so they have to dispose of it and fill the reservoir again.

The ornamental portal of the water reservoir usually shows the wisdom and talent of Iranian architecture. We can see varieties of inscriptions, decorative masonry, etc. on these portals (Figure 1.24).

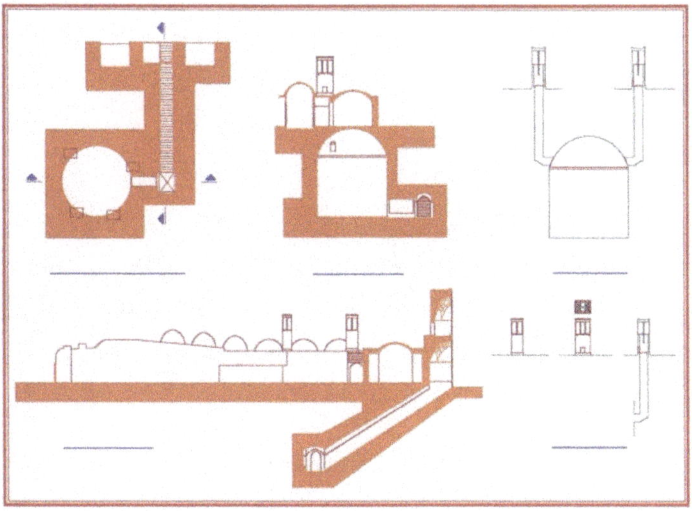

Figure 1.24 Profile and plan of a typical water reservoir.

Water reservoirs were filled in the wet season with the water of a nearby shallow qanat and in the dry season could provide people with cool and fresh water. To prevent sedimentation of the storage tank, a small pool was constructed near the water reservoir and a canal conveyed the qanat's water to the pool. Then after the sediment settled, the clean water in the pool was directed to the storage tank of water reservoir (Semsar and Labbaf, 2017).

1.11 Irrigation Pool

If the flow rate or discharge of the qanat is sufficient, there is no need to store water, and it is transferred to the land directly. But if the discharge is not sufficient, it is then necessary to build a pool near the outlet of the qanat in order to store the qanat water. The water in the pool is then transferred to the farmers' lands (Figure 1.25) for a shorter time, but at a higher flow rate than from the qanat. If there is no pool, there is no possibility for people to use the water to irrigate their lands. Otherwise, the water cannot be transferred to the land efficiently and completely because of a high loss along the way to the cultivated lands (Semsar and Labbaf, 2017).

Figure 1.25 Irrigation pool.

1.12 Bookan

In the absence of vehicles and means of transport in the past, qanat practitioners and workers had to stay away from their houses, near the qanats under construction. So, they had to build temporary houses, named "bookan" (Figure 1.26). The bookan was constructed near the qanat. Its roof was at the same level as the ground surface, and workers entered it by stairs. The thickness of the roof was about 2 meters. Nowadays, bookans like payabs have lost their function, and today there is no active bookan left. But in the past, the bookan was essential, especially for the long qanats; and without it, the work could not be continued. The length of a plain qanat may reach 30 Km and more, so there might be a long distance between workers' house and their work place. It was impossible for the workers to travel this distance and back, every day, so, construction of the bookan was essential for qanat workers. Moreover, the bookan was used as a small forge for sharpening picks. The workers, who rested in the bookan, prepared food for other workers who were digging the qanat and also sharpen the picks for them. The forge in the bookan was connected to the outside through a chimney.

The workers lit the forge and sharpened the picks on it by pounding on them with a sledge hammer. This process was named "kolang keshi". The method of building a bookan was as follows: first the workers dug a slanted gallery in the ground, and at the end of that gallery they built an underground room in a conical shape. They were cautious about probable cave in and so were very careful in constructing a bookan.

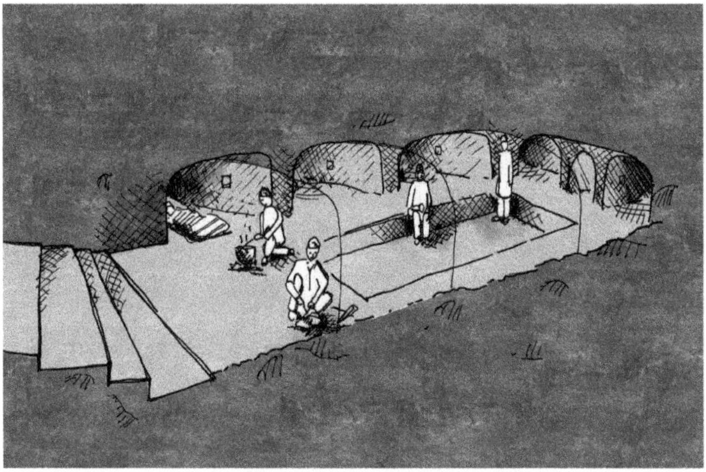

Figure 1.26 Bookan.

To prevent collapse, they used to leave a column of earth intact in the middle of the room. They placed the dirt taken out around the entrance hole to prevent rain from coming in. Another hole was dug to let the smoke of cooking out. Sometimes this hole played the role of a chimney and sometimes an air vent, because air for breathing was also provided by this hole. The size of the bookan depended on the number of the people who were to stay there, the more the people, the bigger the bookan.

Collapse is likely, due to the instability of the earth after leaving the bookan, unless it has been dug in a hard ground.

Depending on the number of muqanis, several niches were dug into the walls, 0.5 meters above the floor, forming small chambers or sleeping platforms. After digging the main parts of the bookan, they also gouge out some smaller niches in the wall of living room, so that the muqanis could put their personal stuff in them; there was also a special place for putting the oil lamps and torches, and another for putting the instruments used for digging. A place separated from the living room was used designed for cooking.

Almost every 3 kilometers along the gallery, a new bookan was built; but if the gallery was dug in a hard area, the muqanis could stay in one bookan for a long time (Semsar and Labbaf, 2017).

Reference

Semsar Yazdi A. A. and Labbaf Khaneiki M. (2017). Qanat Knowledge: Construction and Maintenance, the Netherlands, Springer.

2

History

2.1 Etymology of the Word Qanat

There is controversy about the origin of the word "qanat" among the etymologists. Some consider it a Persian word that has changed into the present pronunciation. They believe that the word "qanat" originated from the word "Kane" which means digging in Persian. In contrast to this theory, some consider an Arabic Origin for the word qanat, since its plural form is qanawat from which the English world canal has emanated. Another name widely used for qanat is "Kariz", "Karez", "Kahrez" or "Kahriz" that is originally Persian, meaning "Straw Pouring"[1].

In the countries benefiting from qanats, different terms are applied to this hydraulic system. (Hum Lum, 1965), (Beaumont, 1971, Bonine, 1989), (Aghasi and Safinejad, 2000). More than 27 terms for qanat are being used in these countries (Figure 2.1):

"Qanat" and "Kariz" in Iran, "Falaj" pl "Aflaj" in Oman, "Kariz" or "Karez" in Afghanistan, Pakistan, Azerbaijan and Turkmenistan, "Ain" in Saudi Arabia, "Kahriz" in Iraq, "Kanerjing" in China, "Foggara" in Algeria,

[1]This name reflects a tradition of water division according to which some straw was poured in the qanat water while distributing water among the farmers, to specify how much water belongs to a particular shareholder.

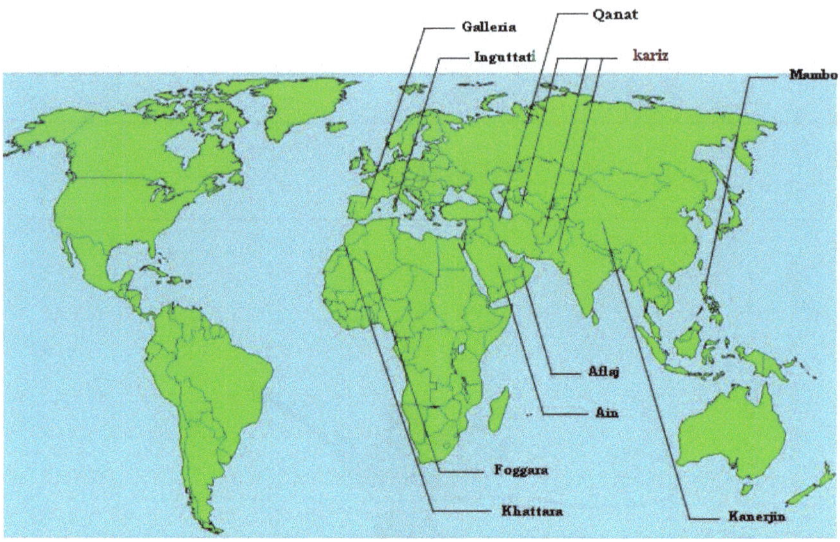

Figure 2.1 Qanat equivalent terms.

"Khattara" or "Khettara" and "Rhettaras" in Morocco, "Galleria" in Spain, "Qanat Romoni" in Syria and Jordan, "Foggara" and "Khettara" and "Iffeli" in North Africa, "Galerias" in the Canary Islands, "Mambo" in Japan, "Inguttati" in Sicily. Some other terms used for qanats are as follows: Ghundat, Kona, Kunut, Kanat, Khad, Konait, Khriga, Fokkara, etc.

2.2 Geographical Distribution of Qanats

The arid and semi-arid regions of the world, whose rainfall shortage does not allow any permanent surface streams, but enjoying rich groundwater resources, have had a good potential to house the system of qanat. So, the system of qanat could be introduced and spread rapidly across such regions.

Iwao Kobori (Kobori, 1964) believes that the qanat system, in all probability, developed in ancient Persia some 2500 years ago and then spread to Afghanistan and eventually along the Silk road as far east as China, as well as by Arabic cultures to the far west including Morocco and Spain (Figure 2.2).

According to Goblot (Goblot, 1979), qanats originated in the northwest of present Iran, dating back to 600–800 B.C. In 525 BC qanats were introduced to the southern coast of the Persian Gulf and then in 500 B.C to Egypt,

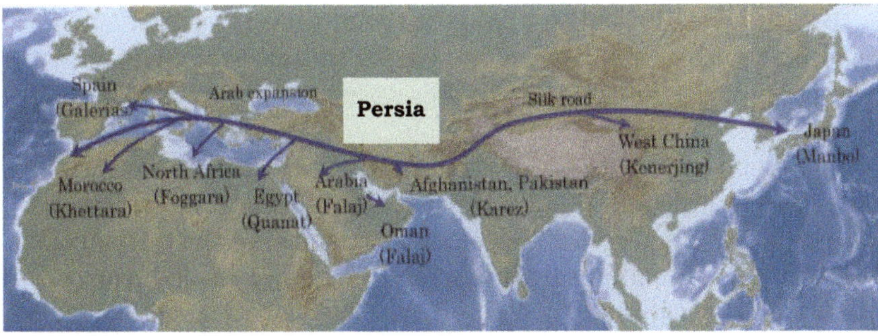

Figure 2.2 Geographical distribution of qanat.

Source: JICA with slight modification.

in 750 AD to Spain, in 850 AD to Southern Algeria, in 1520 AD to Mexico & Los Angeles, in 1540 AD to Pica in Chile and in 1780 AD to Turfan (Northwest of China).

Also, some scholars go far back in antiquity regarding the qanats of Syria. They have maintained that the Hailan-Aleppo qanat, a 12-km long subterranean channel, which functioned until the early part of this century, is coeval with the Aramaeans and their fortress at Aleppo (13th century B.C.) However, this is highly unlikely as the best evidence (archaeological and written accounts) suggests that qanat irrigation was first invented in the Armenian-Persian region about 600–800 B.C. (Lightfoot Dale, 1997).

Jordanians today refer to qanats as “Roman canals”, or qanat Romani (kaneh Romani in northern Jordan). Most scholars believe that Jordan’s qanats were built by the Romans and used by the Byzantines from the 1st century B.C. to the 7th century AD. (Lightfoot Dale, 1996).

The origin and geographical diffusion of qanats have caught the attention of many scholars, and there are some valuable researches on this issue. For example, Haupt a German researcher has studied a qanat system in the area of Lake Van in Turkey (Haupt Lehmann, 1925), and then Weisgerber conducted research on the history of qanats in Oman (Weisgerber Gerd, 2003). Chauveau has studied qanats in the oasis of Kharagha’ in Egypt from archeological point of view (Chauveau Michel, 2001), and Salvini has done some research on the historic hydraulic structures in Urartu (Salvini Mirjo, 2001). Also, there are some scientific reports on the qanats of Oman and Iran, written by Boucharlat, the French archeologist (Boucharlat Remy, 2001). Another scholar, Briant, has scrutinized the text of Polibius on the qanats of Iran (Briant Pierre, 2001). An Iranian historian named Zohreh Cheraghi has provided a valuable anthology on the research conducted on qanat history and its geographical diffusion up to the present in her PhD thesis (Cheraghi Zohreh, 2010).

In a paper published in 2014, Majid Labbaf Khaneiki argues that qanat has been devised as an endogenous collective response of the human settlements to climate change to promote their adaptability to the changing environment. According to him, invention of qanat might take place in different geographical locations independently and simultaneously, though some reliable historical evidence does not let us rule out geographical diffusion of qanat. Nevertheless, it seems that prevailing pattern of genesis of qanat in the ancient world was not "diffusion", but shifting from groundwater to surface water resources in the wake of climate change and then a spontaneous collective response to surface water scarcity seems to be a better answer to the question "how qanat came into existence" (Labbaf Khaneiki, 2014).

The system of the qanat spreads out between the latitudes 15° and 45°N, and this technique can be found even in such rainy regions as Germany. According to some studies, many of the countries between the abovementioned latitudes enjoyed the system of qanat and some of them still profit from this technique: for example, one can mention such Asian countries as Japan, China, Afghanistan, Pakistan (Baluchistan), Iraq (Kurdistan), Oman, Saudi Arabia, Iran, Syria and Azerbaijan (Nakhchivan).

Also, most of the North African countries like Egypt, Libya, Tunisia, Algeria, and Morocco enjoyed the benefit of qanats some of which are still active. In the Canary Islands a considerable number of qanats has been reported. In Europe including Spain, Greece, Germany and Sicily there existed qanats some of which were used as a subterranean canal just to convey water from a surface source to the desired place.

In the new world, the qanat has not played a vital role in supplying water, but some running qanats have been seen in Mexico (Parras), Peru (Nasca), Chile (Pica), Hawaii (Honolulu), USA (California and Los Angeles).

2.3 Historical Evolution of Qanats in Iran

In the course of Iranian history, the qanat has had many ups and downs. Sometimes, the qanats as well as the qanat constructers were supported and encouraged by the government, and sometimes were deserted. Even when the qanats were destroyed for some military purpose, the qanat would start flourishing as soon as the political situation got stable. The risks that threaten qanats today differ from those in the past. In other words, in the past political and military crises had a negative impact on qanats; however, the qanats could recover as soon as the crisis was over. But the present risks are something else, and more destructive in the long term. The present risks are acting environmentally, so it is not so easy to tackle them. Therefore, it is essential for the governments and nations throughout the world to think more about introducing new legislation on the protection of groundwater resources against any kind of overexploitation.

Qanat civilization is rooted in this ancient hydraulic structure. Over the past 3000 years, the system of qanat has underlain many technological, social, moral, economical and legal principles that have formed an important part of the Iranian culture. These principles evolved into the present state by being passed from generation to generation. The present generation should build on these principles, behind which there are three thousand years of history, not to forget about them.

To review the situation of qanats in the course of the Iranian history, we explored some documents on qanats, from the first historical records to the present ones. To do so, two periods – before and after Islam – have been examined. In terms of each period we try to review the situation of qanats keeping pace with the history of kings and governments.

First, it seems necessary to look at the geographical and climatological conditions of Iran, for the natural infrastructure had an important role in creating and developing the qanat systems.

Suffice it to say, Iran has a variable but, in general, arid climate in which most of the relatively scant annual precipitation falls from October through April. In most of the country, yearly precipitation averages 250 millimeters or less. The major exceptions are the higher mountains of the Zagros and the Caspian coastal plain, where precipitation averages at least 500 millimeters annually. In the western part of the Caspian, rainfall exceeds 1000 millimeters annually and is distributed relatively evenly throughout the year. This contrasts with some basins of the Central Plateau that receive 100 millimeters or less of precipitation annually.

2.3.1 Before Islam

It is Henry Goblot who explores the genesis of this technology for the first time. He argues in his book entitled "Qanat; a Technique for Obtaining Water"[2] that during the early first millennium before Christ, for the first time

[2]Les qanats: une technique d'acquisition de l'eau.

some small tribal groups gradually began immigrating to the Iranian plateau where there was less precipitation than in the territories they came from. They came from somewhere with many surface streams, so their agricultural techniques required more water than was available in the Iranian plateau. Hence, they had no option but to fasten their hopes on the rivers and springs that originated in the mountains. They faced two barriers; the first was the seasonal rivers, which had no water during the dry and hot seasons. The second was the springs that drained shallow groundwater and fell dry during the hot season. But they noticed some permanent runoff flowing through the tunnels excavated by the Acadian miners who were in search of copper. These farmers established a relationship with the miners and asked them to dig more tunnels to supply more water. The miners accepted to do that, because there was no technical difficulty for them in constructing more canals. In this manner, the ancient Iranians made use of the water that the miners wished to get rid of it, and founded a basic system named qanat to supply the required water to their farm lands. According to Goblot, this innovation took place in Urartu[3] and later was introduced to the neighboring areas like the Zagros Mountains.

According to an inscription left by Sargon II the king of Assyria, in 714 BC, he invaded the city of Uhlu lying in the northwest of Uroomiye lake that lay in the territory of the Urartu empire, and then he noticed that the occupied area enjoyed a very rich vegetation even though there was no river running across it. Hence, he managed to discover the reason why the area could stay green and realized that there were some qanats behind the matter. In fact, it was Ursa the king of the region who had rescued the people from thirst and turned Uhlu into a prosperous and green land. Goblot believes that the influence of the Medians and Achaemenians made the technology of qanat spread from Urartu (in the western north of Iran and near the present border between Iran and Turkey) to all over the Iranian plateau.

[3]Strictly speaking Urartu is the Assyrian term for a geographical region, while the "Kingdom of Urartu" or the "Biainili lands" are the Iron Age state that arose in that region. That a distinction should be made between the geographical and the political entity was already pointed out by König. The landscape corresponds to the mountainous plateau between Asia Minor, Mesopotamia, and the Caucasus mountains, later known as the Armenian Highlands. The kingdom rose to power in the mid 9th century BC and was conquered by Media in the early 6th century BC. (http://en.wikipedia.org/wiki/Urartu)

Although Goblot's theory may be valid in Urartu or neighboring regions, we believe that we should be more realistic toward the invention of qanat. It is more likely that the first qanats were built in the central plateau of Iran at the mountain bases or along the valleys, though we do not rule out the possibility of construction of qanats outside of this region for example in Oman or Urartu independently. But it is hard to accept the theory that the qanat was first invented by the copper miners in Urartu and then introduced to the Iranian plateau and used by the farmers who lived some 1500 kilometers away from its origin. Not only in the past, but also at present, the immediate reaction of any farmer is to dig into a spring when its water dwindles. In the mountainous region surrounding the Iranian desert there were many natural springs, which supplied water to the small communities who lived there. In the wake of climate change, the precipitation reduced and accordingly many of the springs dried up or just trickled. In this situation the immediate reaction of the people might be to dig the same springs to track the water, and after a while they ended up building a long tunnel with some shaft wells through which they could better haul the debris to the surface. In fact, we consider that the natural springs led the people to construct the first qanats, and it is very likely that an ancient man would be inspired by a trickling spring to burrow back to get closer to the source of water. Probably that was how the system of qanat came into existence, probably in several regions simultaneously. Hasanalian, who has conducted much research on Sialk,[4] has come to the conclusion that the spring of Fin which once provided this ancient settlement with water was later manipulated and turned into a qanat (Hasanalian Davood, 2006). Even a few years ago, we witnessed this process in an off the beaten path village in southern Khorasan. In this village, there was a natural spring with a discharge of about 3 liters per second. After a drought broke out some 15 years ago, the water of this spring dramatically decreased and as a result the villagers made up their mind to deepen the spring to reach water again. They dug the spring horizontally up to 30 meters and every year they extended this tunnel to keep the discharge steady. After 20 years that spring was turned into a qanat with two shaft wells. This scenario can be seen beside the theory that the miners of Urartu invented the qanat as a byproduct, and could be repeated wherever enjoyed suitable conditions for qanats were present. Who knows how or even whether the farmers in the central Iran came in contact

[4] Sialk is a large ancient archeological site near the city of Kashan, in central Iran, tucked away in the suburbs, close to Fin Garden. This civilization dates back to 5500 BC (Fazeli et al. 2010).

with those miners in Urartu and how they learnt this technique and how they brought it to the central Iran (Labbaf Khaneiki, 2014)?

2.3.2 Achaemenian Empire (550–330 BC)

It was an Achaemenian official ruling that in case someone succeeded in constructing a qanat and bringing groundwater to the surface in order to cultivate land, or in renovating an abandoned qanat, the tax he was supposed to pay the government would be waived not only for him but also for his successors for up to 5 generations. During this period, the technology of qanat was in its heyday and it even spread to other countries. For example, following Darius's order, Silaks the naval commander of the Persian army and Khenombir the royal architect managed to construct a qanat in the oasis of Kharagha' in Egypt (Figure 2.3). "Beadnell" believes that qanat construction dates to two distinct periods. In Egypt some qanats were constructed by the Persians for the first time, and later Romans dug other qanats during their reign in Egypt from 30 BC to 395 AD. In any events the magnificent temple built in this area during Darius's reign shows that there was a considerable population depending on the water of qanats. Ragerz has estimated this population to be 10,000 people.

Boucharlat confirms the introduction of the qanat to Egypt at the time of the Achaemenids, though he underlines the differences between them and the original Persian qanats. He writes that "recent excavations both in Egypt and Libya have provided evidence of underground galleries dated to the Persian period (mid-fifth century B.C.). However, it should be stressed that these galleries tap water in a very particular geological context, i.e. the huge Nubian Reservoir, a fossil aquifer located in the Nubian sandstone plateau. While the galleries at Ain Manawir also in the Khargeh [Kharagha'] oasis are confidently dated to 4th–5th centuries B.C. and onwards, they are of a type which presents certain specific characteristics. They fit the true qanat-falaj definition in the specific meaning I give to this technique, i.e. draining water from an aquifer (one of the mother wells is 30 m deep). However, they differ from the typical qanat as well: they benefit from artesian sources; they have very short galleries (less than 400 m long, often 200 m); and they have a gradient superior to 10.7% (up to 13%). Tentatively I would suggest that the type of deep gallery used in Egypt represents an adaptation to a local context, but it would not have spread elsewhere because of its specific context" (Boucharlat Remy, 2001).

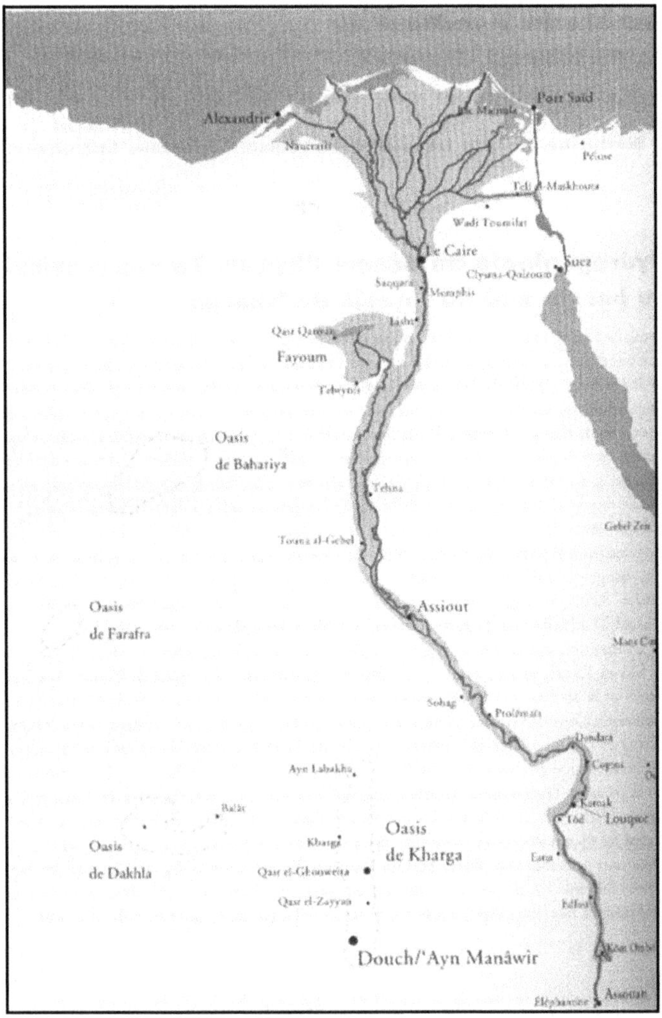

Figure 2.3 Oasis of Kharga (Kharagha) in Egypt.

The most reliable document confirming the existence of qanats in the Achaemenid period has been written by Polibius who states that: "the streams are running down from everywhere at the base of Alborz mountain, and people have transferred too much water from a long distance through some subterranean canals by spending much cost and labor".

2.3.2.1 Seleucid era (312–250 BC)

During the Seleucidian era, which began after the occupation of Iran by Alexander, it seems that the qanats were abandoned. Maybe this situation has something to do with the origin of the occupiers who had no idea on how the qanat system works and should be treated.

2.3.2.2 Parthian era (250 BC–150 AD)

In terms of the situation of qanats during this era, some historical records have been found. In a study by Russian orientalist scholars it has been mentioned that the Persians used the side branches of rivers, mountain springs, wells and qanats to supply water. The subterranean galleries excavated to obtain groundwater were named as qanat. These galleries were linked to the surface through some vertical shafts, which were sunk to get access to the gallery to repair it if necessary.

According to historical records, the Parthian kings did not care about the qanats the way the Achaemenid kings and even Sassanid kings did. Polybius [Greek historian, 203–120 B.C.] records how Arsac III one of the Parthian kings tried to demolish the qanats and so cut off the water supply in order to halt the advance of Antiochus towards the lost Parthian capital of Hecatompylos. Arsac destroyed some qanats to make it difficult for Seleucidian Antiochus to advance further while fighting him (Beaumont, 1971).

2.3.2.3 Sassanid era (226–650 AD)

The historical records from this time indicate a perfect regulation on both water distribution and farmlands. All the water rights were recorded in a special document, which was referred to in case of any transaction. The lists of farmlands – whether private or governmental – were kept at the tax department. During this period there were some official rulings on qanats, streams, construction of dams, operation and maintenance of qanats, etc. The government proceeded to repair or dredge the qanats that were abandoned or destroyed for any reason and construct the qanats if necessary. A document written in the Pahlavi language[5] pointed out the important role of qanats in developing the cities at that time.

[5]An ancient branch of the Persian language that was spoken during the Sassanid era.

2.3.3 After Islam (621–1921 AD)

In Iran, the advent of Islam which coincided with the overthrow of the Sassanid dynasty brought about a profound change in religious, political, social and cultural structures. But the qanats stayed intact, because the economic infrastructures including qanats were of great importance to the new government. As an instance, M. Lombard reports that the Moslem clerics who lived during the Abbasid period, such as Abooyoosef Ya'qoob (death 798 AD) stipulated that whoever can bring water to the idle lands in order to cultivate them, his tax would be waived and he would be entitled to the lands cultivated. Therefore, this policy did not differ from that of Achaemenids not taking any tax from the people who revived abandoned lands. The Arabs' supportive policy on qanats was so successful that even the holy city of Mecca gained a qanat too. The Persian historian Hamdollah Mostowfi writes: "Zobeyde Khatoon (Haroon al-Rashid's wife) constructed a qanat in Mecca. After the time of Haroon al-Rashid, during the caliph Moghtader's reign this qanat fell into decay, but he rehabilitated it, and the qanat was rehabilitated again after it collapsed during the reign of two other caliphs named as Ghaem and Naser. After the era of the caliphs this qanat completely fell into ruin because the desert sand filled it up, and later Amir Choopan repaired the qanat and made it flow again in Mecca."

There are also other historical texts proving that the Abbasids were concerned about qanats. For example, according to the "Incidents of Abdollah bin Tahir's Time" written by Gardizi, in the year 830 AD a terrible earthquake struck the town of Forghaneh and reduced many homes to rubble. The inhabitants of Neyshaboor used to come to Abdollah bin Tahir in order to request him to intervene, for they bickered over their qanats and found the relevant instruction or law on qanat as a solution neither in the prophet's quotations nor in the clerics' writings. So Abdollah bin Tahir managed to bring together all the clergymen from throughout Khorasan and Iraq to compile a book entitled "Alghani" (The Book of Qanat). This book collected all the rulings on qanats, which could be of use to whoever wanted to judge a dispute over this issue. Gardizi added that this book was still applicable to his time, and everyone made references to this book.

One can deduce from these facts that during the abovementioned period the number of qanats was so considerable that the authorities were prompted to put together some legal instructions concerning them. Also, it shows that from the ninth to eleventh centuries the qanats that were the hub of the agricultural systems were also of interest to the government. Apart from

The Book of Alghani, which is considered as a law booklet focusing on qanat-related rulings based on Islamic principles, there is another book about groundwater written by Karaji[6] in the year 1010. This book entitled Extraction of Hidden Waters examines just the technical issues associated with the qanat and tries to answer the common questions such as how to construct and repair a qanat, how to find a groundwater supply, how to do leveling, etc. some of the innovations described in this book were introduced for the first time in the history of hydrogeology, and some of its technical methods are still valid and can be applied in qanat construction. The content of this book implies that its writer (Karaji) did not have any idea that there was another book on qanats compiled by the clergymen. Mohammad bin Hasan quotes Aboo-Hanifeh that in case someone constructs a qanat in abandoned land, someone else can dig another qanat in the same land on the condition that the second qanat is 500 zera' (375 meters) away from the first one.

Ms. Lambton quotes Moeen al-din Esfarzi who wrote the book Rowzat al-Jannat (the garden of paradise) that Abdollah bin Tahir (from the Taherian dynasty) and Ismaeel Ahmed Samani (from the Samani dynasty) had several qanats constructed in Neyshaboor. Later in the 11th century, a writer named Nasir Khosrow acknowledged all those qanats by the following words: "Neyshaboor is located in a vast plain at a distance of 40 Farsang (~240 km) from Serakhs and 70 Farsang (~420 km) from Mary (Merv) ... all the qanats of this city run underground, and it is said that a traveler who was offended by the people of Neyshaboor has complained that; "what a beautiful city Neyshaboor could become if its qanats would flow on the ground surface and instead its people would live underground".

These documents all certify the importance of qanats during Islamic history within the cultural territories of Iran.

In the 13th century, the invasion of Iran by Mongolian tribes reduced many qanats and irrigational systems to ruin, and many qanats were deserted and dried up. Later in the era of Ilkhanid dynasty, especially at the time of

[6]Karaji, Mohammed ibn al-Hasan al-Hasib (early 10th century AD). The advanced knowledge of the Persians concerning groundwater is demonstrated by his book entitled: The Extraction of Hidden Waters. It reveals a profound and exacting technical understanding of groundwater theory and is the oldest known text on the subject. Karaji's knowledge of groundwater is in general agreement with modern understanding of the subject. For example, he was familiar with the general concepts of the hydrological cycle. While he never featured the whole cycle as we know it, he records in different passages of his book each individual phase. Karaji spent most of his life in Bagdad, working mainly as a mathematician. He wrote several books on algebra and geometry which were translated into German during the 19th century. (Pazwash, Hormoz and Gus Mavrigian, 1980) – also too long!

Ghazan Khan and his Persian minister Rashid Fazl-Allah, some measures were taken to revive the qanats and irrigational systems. There is a 14th century book entitled Al-Vaghfiya Al-Rashidiya (Rashid's Deeds of Endowment) that names all the properties located in Yazd, Shiraz, Maraghe, Tabriz, Isfahan and Mowsel that Rashid Fazl-Allah has donated to the public or religious places. This book mentions many qanats running at that time and irrigating a considerable area of farmlands. At the same time, another book entitled Jame' al-Kheyrat was written by Seyyed Rokn al-Din on the same subject as Rashid's book. In this book Seyyed Rokn al-Din names the properties donated in the region of Yazd. These deeds of endowment indicate that much attention was given to the qanats during the reign of Ilkhanids, but it is attributable to their Persian ministers who influenced them.

In the Safavid era (15th and 16th centuries) the problem of water shortage intensified and led to the construction of many water reservoirs and qanats. Chardin, the French explorer who made two long journeys to Iran at this time reports that: "the Iranians rip the foothills in search of water, and when they find any, by means of qanats they transfer this water to a distance of 50 or 60 kilometers or sometimes further downstream. No nation in the world can compete with the Iranians in recovering and transferring groundwater. They make use of groundwater in irrigating their farmlands, and they construct qanats almost everywhere and always succeed in extracting groundwater."

The dynasty of Qajar ruled Iran from the 16th century to the early 18th century. According to Goblot, the time of Qajar can be considered as the heyday of qanats, for the qanats could flourish. Agha Mohammad Khan the founder of Qajar dynasty chose Tehran as his capital city, a city where there was no access to a reliable stream of surface water and it had to rely on groundwater. The rich supply of groundwater and suitable geological-topographical conditions of Tehran allowed this city to house many qanats, whose total discharge amounted to 2000 liters per second. Haj Mirza Aghasi (ruling between 1834 and 1848), the prime minister of the third king of the Qajar dynasty encouraged and supported qanat construction throughout the country[7]. Jaubert de Passa who has surveyed the situation of irrigation in Iran reports a population of 50,000 in Hamedan, 200,000 in Isfahan and 130,000 in

[7] According to a famous story, one day Haj Mirza Aghasi paid a visit to a qanat to find out how they are getting on their work. He asked the worker who was at the bottom of a well if the qanat has got to the water or not. The worker who did not recognize the prime minister complained that Haj Mirza Aghasi is wasting the country's budget on qanats that will never have water. The minister replied: "don't worry man! If the qanat will not get us water, it will get you a living instead". The minister's word has turned into a popular proverb in Iran.

Tehran in the year 1840. Then he claims that in these cities, life is indebted to the qanats, which are being constructed in a simple but powerful manner. In a nutshell, the period of Qajar that lasted about 1.5 centuries witnessed the considerable endeavors to revive and build new qanats.

2.3.4 Period of Pahlavi

The period of Pahlavi began in 1921 when Reza Shah came to power and ended in 1979 with his son's overthrow in the wake of the Islamic revolution. the process of qanat construction and maintenance continued during this time, but overall the materialist trends inflicted some damage on this irrigation system. Unfortunately, most Iranian scholars had a low opinion of traditional technology during the Pahlavi period. They assumed that the only way out of economic crisis is to belittle traditional methods to pave the way for modern technologies. Some of the scholars and politicians even tried to exaggerate the technical defects in traditional methods to convince the public to get rid of them.

In the face of such a paradigm, a department that was responsible for the qanats was set up by Reza Shah. At that time most of the qanats belonged to the landlords. In fact, feudalism was the prevailing system in the rural regions. The peasants were not entitled to the lands they worked on but were considered just as the users of the lands. They had to pay rent for land and water to the landowners. These people, who were relatively wealthy, could afford to finance all the activities required to maintain the qanats. According to the report of Safi Asfiya, who oversaw supervising the qanats of Iran in the former regime, in the year 1942 Iran had 40,000 qanats with a total recharge of 600,000 liters per second or 18.2 billion cubic meters per year.

This period faced the advent of the new technology of pumped well, which was unconditionally welcomed by the government on one hand, and experienced a reformative program declared by the former Shah on the other hand. These two issues, which are described in the next part of this chapter, played an important role in phasing out of qanat systems.

In 1946, the Independent Foundation on Irrigation was set up by the government for the first time, aiming at the construction of dams and meteorological stations, implementation of irrigation projects, supplying drinking water and harvesting surface runoff. According to a report published by this foundation in 1954, the first pumped well was drilled in Sanglaj followed by several wells sunk in other districts of Tehran. Later, many other wells were drilled in Tehran, Kerman, Qom, Abadan, Qaen, etc. to supply drinking water to the residents. As mentioned, the first pumped wells were drilled in early 1950s when many of the land owners considered pumped wells as a means for developing their farm lands. At that time, there was no legislation in terms of drilling pumped wells and nobody could be put on trial for such a charge unless the owners of the nearby qanats would file a complaint about the negative impact of a particular well on their qanats. But it would take a long time to process such complaints, and the owners of the pumped wells, who were wealthier and at a higher economic level usually won the argument. Suffice it to say, widespread conflict between users of pumped wells and the qanats broke out, whose final loser was the qanats.

In 1959 a reformative program named as White Revolution was declared by the former Shah. One of the articles of this program addressed land reform[8] that let the peasants take ownership of a part of the feudal lands. In fact, the land reform dashed the landlords' hope. They lost their motivation for investing more money in constructing or repairing the qanats, which were subject to the land reform law. In addition, the peasants could not come up with the money to maintain the qanats, so many more qanats were gradually deserted. The introduction of modern devices that made it possible to drill many deep wells and extract groundwater much more quickly and easily aggravated the annihilation of the qanats. The pumped wells had a negative impact on the qanats due to their overexploitation of groundwater. These changes that occurred in Mohammad Reza Shah's reign inflicted such great damage on the qanats of the country that many vanished forever. In 1961, another report was published revealing that in Iran there were 30,000 qanats out of which just 20,000 were still in use with a total output of 560,000 lit/se or 17.3 billion cubic meters per year. The statistics related to 14,778 qanats estimate their overall discharge 6.2 billion cubic meters per year between 1972 and 1973. If we assume the total number of the qanats at that time to be 32,000, their annual discharge could have amounted to 12 billion cubic meters.

In 1963, the Ministry of Water and Electricity was established to provide the rural and urban areas of the country with the sufficient water and electricity. Later this ministry was renamed as the Ministry of Energy. Three years later in 1966 the parliament passed a law protecting groundwater resources. According to this law, the Ministry of Water and Electricity was allowed to

[8]The issue of land reform in Iran has been described in detail in the Chapter 3.

ban drilling any deep and semi-deep wells[9] wherever surveys showed that the water table was dropping because of over-pumping. In fact, this law was passed when the growing number of the pumped wells sounded the alarm about the over-pumping and depletion of groundwater leading to a decline in qanat flow all over the country. This law as well as the law of water nationalization that was approved in 1968 and eventually the law of fair distribution of water passed (in 1981) after the Islamic revolution emphasized the definition of restricted and free areas for drilling. In the restricted areas, drilling any wells (except for drinking and industry) was prohibited to prevent the continuous depletion of groundwater, so that the other qanats had a better chance to survive.

One of the advantages of such legislation was to better regulate the exploitation of groundwater. The farmers gained an understanding of the fact that whoever wanted to drill a new well, deepen or replace his well should get the permission from the ministry of energy. They knew that nobody has the right to pump groundwater resources all they want, but they can withdraw only as much water as their permit allows. Also, they understood that all these limitations are to protect groundwater resources against over pumping and eventually work to the advantage of those who are using these resources. In fact, in many of the arid parts of the country, groundwater resources were heading into crisis and the depletion of water table was continuing, while illegal pumped wells still mushroomed all over the country. Especially at the end of the former Shah's time, between 1978 and 1979, the government's weakness in preventing the violations of law led to a growth in the number of illegal wells. Also, some parts of the national pasture lands were turned into farm lands. The reasons for such illegal acts, which are rooted in economic needs of some farmers and the wrong policies of that time should be surveyed in a separate research by social scientists and economists.

2.3.5 Time of the Islamic Republic

After the Islamic revolution, special attention was given to the qanats. For the first time in 1981 a conference on qanats was held in Mashhad during which the different options to mitigate the problem of overexploitation

[9]According to the definitions of the Iranian water industry standard, deep well is sunk some meters below water table and is drilled by machinery, and in some cases a deep well may intercept one or more confined or unconfined aquifers, whereas a semi deep well may be dug by machinery or by hand and discharges water only from the upper parts of unconfined aquifers.

of groundwater were explored. The organization of Jahad Sazandegi took responsibility for the rehabilitation of qanats and subsidized their shareholders. Now the same organization which was renamed as Ministry of Jahad Agriculture is responsible for the qanats and continues to grant some funds to the stakeholders to maintain their qanats. During recent years (from 2005 to 2009), parliament has allocated an annual budget of 15 million USD to this ministry for construction and maintenance of the qanats[10]. Many other qanats would dry up without this budget, because the owners cannot afford to pay all the expenses. In most cases, 50 percent of the budget needed for a qanat is supplied by the government and the rest by the owners themselves. Moreover, in the past, a considerable budget went to the qanat rehabilitation. From 2000 to 2010, a budget of 886,300,000,000 USD was spent on the renovation of 22,356 qanats in Iran, and this project resulted in an increase of 2018 million cubic meters of qanat water (Parastar Alireza, 2010).

In the years 1984–1985, the ministry of energy held a census of 28,038 qanats whose total annual discharge was 9 billion cubic meters. In the years 1992–1993, a census of 28,054 qanats showed a total discharge of 10 billion cubic meters. Ten years later in 2002–2003 the number of the qanats was reported to be 33,691 with a total discharge of 8 billion cubic meters.

In the year 2000, holding an International conference on qanats in Yazd drew a lot of attention to the qanats. In 2005 the Iranian government and UNESCO signed an agreement to set up the International Center on Qanats and Historic Hydraulic Structures under the auspices of UNESCO.

The main mission of this center is on recognition, transfer of knowledge and experience, promotion of information and capacities regarding all the aspects of qanat technology and other historic hydraulic structures. This mission is aimed at achieving sustainable development of water resources and the application of the outcome of the activities to preserve historical and cultural values as well as the promotion of the public welfare within the communities whose existence depends on the rational exploitation of the resources and preservation of such historical structures. Another mission is to promote research and development to restore the qanats and other traditional historic hydraulic structures for sustainable development, through international co-operation and global transfer of knowledge and technology.

[10]According to the director of the qanat department of Jahad agriculture ministry, for the next five years (national fifth developmental program; 2009–2014) this budget will increase to 30 million USD annually.

In 1999, a scientific resolution entitled Water Allocation System was declared by the ministry of energy, according to which allocating water (whether underground or surface) to any sector (whether agricultural or industrial or domestic) should be in proportion to the hydrological balance in a particular basin. This resolution was later, in 2003, modified, and is still valid as a basis of action. All these resolutions are to the advantage of qanats, since they control the exploitation of groundwater reserves. At present, each regional water authority that manages some particular basins should calculate the water balance every year to know the ratio of inflow to outflow. In case the amount of inflow prevails over the amount of outflow, the authority can issue some permits to discharge water while giving priority to the domestic and industrial sectors. If there is any water left, it goes to the agricultural sector. This resolution, which is very comprehensive and well developed, is in fact a system of water accounting, being able to inform the authorities and decision makers about the water balance in their own regions. Under present conditions, the first step to construct a dam is to conduct research on the water shares supposed to be allocated. The water built up behind a given dam should suffice to irrigate the downstream lands and satisfy environmental needs; otherwise the dam construction project would be halted at this stage.

On October 19, 2003, the Iranian governmental council defined its long-term development strategies for Iran's water resources:

The valuable document "Long-term Development Strategies for Iran's Water Resources" will be a suitable guide to compile the medium- and short-term plans of national Water management, resulting in the optimized exploitation of national Water Resources by uniting all the arenas of Water management.

Long-term Development Strategies for Iran's Water Resources is based on coordination among various sectors, planning to observe water resources capacity on the basis of sustainable development principles and meet the needs of irrigated lands within basins, water consumption pattern reform and water losses management, the adaptation of economic development programs to the development projects in each basin, taking into account the economic and environmental value of water, water resources pollution management and control, reasonable and efficient pricing of water, observing national interests, natural and social rights considering the projects of transferring water between basins and water exchanges, the decentralization of national water resources management structure, the interested parties' participation to adapt to provincial development projects and to prepare and carry out risk management (droughts and floods) and public awareness programs, statistics of joint water resources flowing out, equipping and completing water gauging networks and conserving, reviving and operating historical water structures are among the measures foreseen and compiled in the document. The document approved by the Cabinet, being a national necessity, is regarded as a pioneering measure for international policies" (Ardakanian Reza, 2003).

In a nutshell, currently, the Iranian government has sufficient legislative and organizational background to operate integrated water management. The lawlessness regarding water exploitation has dropped to a minimum. Illegal deep wells are rarely drilled, and the existing pumps are no longer replaced with the stronger ones. But some arid parts of the country are still suffering from the environmental problems of past overexploitation. Between 1977 and 1997 the number of the pumped wells reached a peak, but later started dropping. This means that the pumped wells are under control. At the present, the total discharge of the qanats is almost steady, and this fact represents how concerned the government is about the preservation of this valuable treasure.

According to a report published in 2005 by the Water Resources Base Studies Department affiliated to the ministry of energy, there are 15 Regional Water Authorities throughout the country[11], based in 30 provinces, conducting research projects on water resources in 609 study sites. Note that each study site is the smallest research unit, containing one or several catchments (Figure 2.4).

Out of the 609 study sites, 214 sites with an overall area of 991, 256 square kilometers have been declared as restricted regions, and 395 sites with an area of 630, 648 square kilometers are considered free. In the restricted regions there are 317, 225 wells, qanats, and springs that totally discharge 36,719 million cubic meters water a year, of which 3,409 million cubic meters are more than the aquifer capacity. This deficit in the volume of the aquifer reserves has led to a long-term groundwater level drop of 41 centimeters a year, on average. In the free regions the number of wells, qanats and springs amounts to 241,091 with an output of 37,527 million cubic meters a year. Therefore, in 2005, in all over the country, there were 130,008 deep wells with a discharge of 31,403 million cubic meters, 338,041 semi deep wells with a discharge of 13,491 million cubic meters, 34,355 qanats with a discharge of 8,212 million cubic meters, and 55,912 natural springs with a discharge of 21,240 million cubic meters (Figure 2.5).

---

[11] At the time of this report, there were 15 regional water authorities, but later their number increased to 30, one for each of the 30 provinces of Iran.

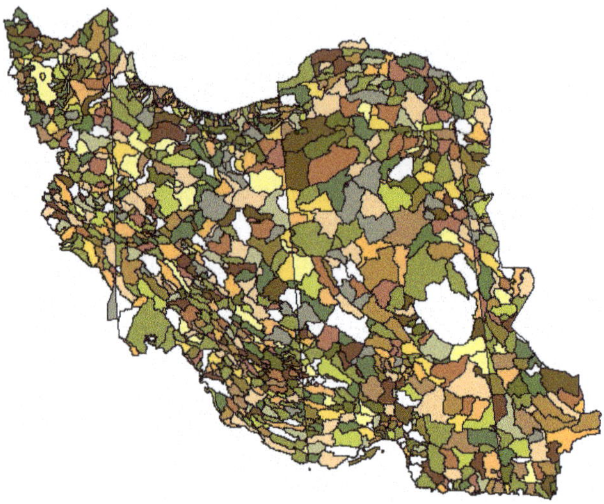

Figure 2.4 Study sites in Iran based on watersheds.

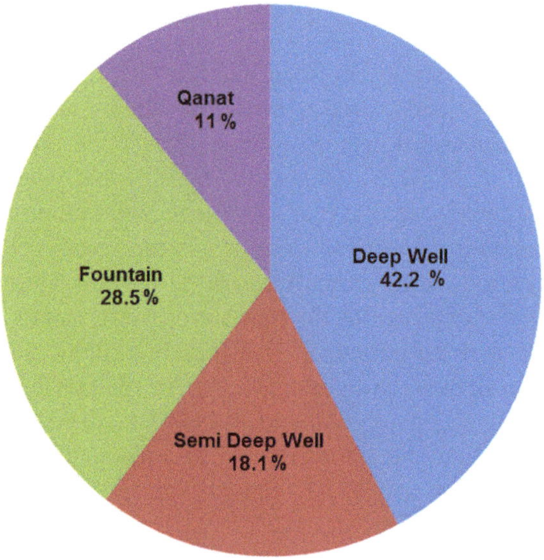

Figure 2.5 Percentage of each water source out of the total groundwater discharge in Iran in 2005.

A recent census conducted by the Iranian ministry of energy shows a rise in the number of qanats in the face of a decline in their total discharge. According to this census, the qanats number 41,031 with a total discharge of 4,531 million cubic meters. Such a dramatic decline in the discharge of qanats contradicts an increase in the number of qanats. The increase in the number of qanats does not suggest that new qanats have been built in the country, but it is down to the nature of mountainous short qanats whose discharges closely correlate with fluctuations in rainfall. Such qanats might not be included as active qanats in the 2005 census. The 2018 census reports the number of deep wells to be 210,784 with a total discharge of 33,423 million cubic meters per year, 621,910 semi-deep wells with a discharge of 12,678, and 173,474 natural springs with a discharge of 10,684 million cubic meters (Iran Water Resources Management Company, 2018; Figure 2.6).

In fact, the Iranian water authorities may succeed in balancing groundwater discharge with recharge in the restricted regions and prevent the free regions from turning into the restricted ones if: (1) The regional water authorities keep up their policy on controlling withdrawal of groundwater resources. (2) A campaign is started on enhancing public awareness about groundwater preservation. (3) The irrigation efficiency is improved, and suitable cropping

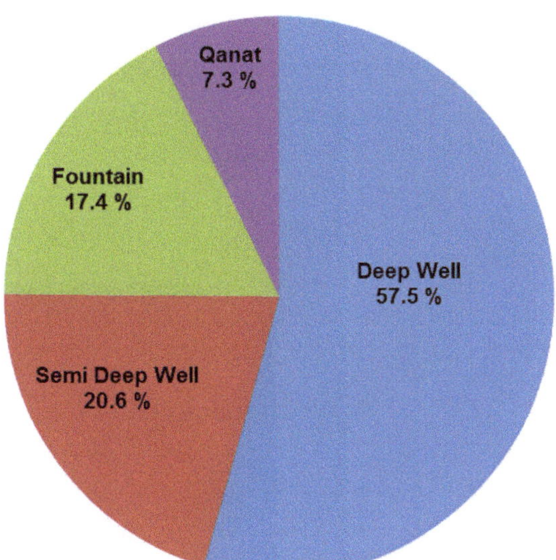

Figure 2.6 Percentage of each water source out of the total groundwater discharge in Iran in 2018.

patterns are introduced in keeping with the different environmental conditions. For example, in the arid and semi-arid regions, the farmers should be prevented from cultivating crops consuming much water, such as vegetables, and instead be encouraged to produce such crops as olives or pistachio, which need less water.

Given the well -developed water laws and rules practiced in the Islamic Republic of Iran as well as the presence of the Regional Water Authorities that are in charge of water management, stopping groundwater depletion is not beyond hope.

To mitigate the over-exploitation crisis and to minimize their long-term environmental and ecological consequences, a long-term comprehensive plan has been launched in 2014 by Iranian Ministry of Energy. This integrated plan

Figure 2.7 Selection of Iranian qanats based on their unique characteristics for UNESCO World Heritage List.

pertains to different programs such as society, institution, practice and legal aspects. For every dimension a series of related programs and activities are considered. For example, in terms of society, the plan emphasizes that all the levels should be trained and involved in water resources affairs given the potential role they can play in water resources management. This educational program should begin from the first level or primary schools to the highest levels even to an administrative scale. The water consumers who are mostly the farmers should receive a continuous and purposeful training program in order to stay informed on the situation of water resources (Semsar Yazdi, 2013).

Iranian qanat was nominated for the UNESCO World Heritage List. Many Iranian qanats bear some characteristics attesting to feats of engineering, considering the intricate techniques used in their construction. For example, there are qanats with extremely deep wells, and others have a two-storey tunnel. Each qanat has unique characteristics that deserve attention. These characteristics were considered while having nominated some qanats for UNESCO World Heritage List. The figure below shows the geographical distribution of selected qanats across the Iranian central plateau, which were placed on UNESCO World Heritage List in 2016 (Semsar Yazdi, Labbaf Khaneiki, 2018; Figure 2.7).

References

Aghasi and Safinejad. (2000). Qanat Glossary, Yazd Regional Water Authority publications, Iran.

Ardakanian, Reza, (2003). Long Term Development Strategies for Iran's Water Resources, Public Relations and International Affairs bureau of Iran water resources management organization, p. 3–4.

Beaumont P. (1971). Qanat systems in Iran. Bulletin of the International Association of Scientific Hydrology, 16(1).

Boucharlat, Remy, (2001). Les galleries de captage dans la peninsula d'Oman au premiere millenaire avant j.c: questions sur leur relations avec les galleries du plateau Iranien, Paris, in: Briant, Persika 2.

Boucharlat, Remy, (2001). Iron Age Water-draining Galleries and the Iranian Qanat, proceedings of the first international conference on the archeology of the U.A.E., p. 170.

Briant, Pierre, (2001). Polybe et les qanates, Paris, seminaire tenu au colloge de France, Persika 2.

Chauveau, Michel, (2001). Les qanates dans les ostraca de Manawir, Paris, Seminaire tenu au au colloge de France, Persika 2.

Cheraghi, Zohreh, (2010). Role of Qanats in the Historic Life of Yazd from Ilkhanid Dynasty to Pahlavi I Era (1295–1941); a critical study, Isfahan University.

Fazeli, H., Beshkani A., Markosian A., Ilkani H., Young R. L., (2010). The Neolithic to Chalcolithic Transition in the Qazvin Plain, Iran: Chronology and Subsistence Strategies: in Archäologische Mitteilungen Aus Iran and Turan 41, pp. 1–17.

Goblot H., (1979). Les Qanats, une technique d'acquisition de l'eau/English: Qanat a techniquefor obtaining water, Paris-La Haye, Mouton/Ecole des hautes en sciences sociales, 236 P. Translated from French to Persian by A. Sarvqad Moqadam, M. H. Papoli Yazdi, 1992.

Hapt, Lehmann, (1925). Armenien einst und yetst, Leipzig.

Hasanalian Davood, (2006). Sialk; the Unsolved Puzzle of Archaeology, Iran Newspaper Archive at http://www.magiran.com/npview.asp?ID=1295401

Iran Water Resources Management Company, (2018). Basic Study on Water Resources, Ministry of Energy, Unpublished.

Kobori I., (1964). Some Considerations on the Origin of the Qanat System, in memorial collected papers dedicated to Prof. E. Ishida, Tokyo.

Labbaf Khaneiki, M., (2014). Genesis of Qanat as a Collective Response to Climate Change in Iran (in Persian). Asar Scientific Journal. Iranian Cultural Heritage Organization. No. 66, fall 2014.

Lightfoot Dale R. (1997). Qanats in the Levant: Hydraulic Technology at the Periphery of Early Empires. Technology and Culture 38(2): 432–451.

Lightfoot Dale R. (1996). Syrian Qanat Romani: History, Ecology, Abandonment. Journal of Arid Environments 33(3): pp. 321–336.

Parastar, Alireza, (2010). Unpublished report of office for agricultural water management, Iranian ministry of agriculture.

Pazwash, Hormoz and Gus Mavrigian, (1980). "A Historical Jewelpiece-Discovery of the Millennium Hydrological Works of Karaji, Water Resources Bulletin, December, pp. 1094–1096.

Salvini, Mirjo, (2001). Pas de Qanat en Urartu, Paris, Irrigation et drainage dans l'antiquite, qanats et canalisation souteraines en Iran, Egypt et en Grece.

Semsar Yazdi Ali Asghar, (2013). How to preserve groundwater in arid lands? Proceedings of the International conference on Water, Environment, Energy and Society, 8–11 May 2013, Tunisia.

Semsar Yazdi, A., Labbaf Khaneiki, M., (2018). Qanâts of Iran: Sustainable water supply system. In: Water and Society from Ancient Times to the Present. Sulas, F., Pikirayi, I. (Eds.) England. Routledge.

Weisgerber, Gerd, (2003). The impact of the dynamics of Qanats and Aflaj on oases in Oman: comparisons with Iran and Bahrain, Luxemburg, Internationales frontinus-symposium.

3

Social Aspects

3.1 Qanat-Based Civilization

We should see the qanat as a technique that underlies social, economic and political structures in the central plateau of Iran, whose annual precipitation is so little that the people living there have no choice but to resort to extracting groundwater, mainly through the qanat system. We believe that from a historical standpoint, civilization in Iranian territories is anchored in the surface streams, which once flowed, even in the central plateau. Maybe it is hard for a visitor to believe that most of the Iranian desert once enjoyed permanent rivers, which made a quite different landscape from what we see today; fresh water streams flowing down the foot of the mountains toward the central lakes. There are some archeological sites in the Iranian desert that bear witness to the fact that this area was once populated by people who lived off surface streams, which are replaced by mirages now. For example, there is a long gully near Gonabad in arid southern Khorasan. In years when the seasonal sporadic rains pour down on the surrounding mountains, temporary runoff flows down this gully to the central desert. This water is now so salty that no living thing can use it, and even the most resistant plants to salinity cannot feed on this water. Tens of meters from both sides of the gully are encrusted by salt and completely barren. On the bank of this dry and salty gully, some relics of ancient human settlements take us by surprise, and immediately the question flashes through our mind of how these people could cope with this harsh environment where there is no surface fresh water and the topography does not allow the building of a qanat. To answer this question, we should travel back through time to when this area enjoyed a different climate. We can find many ancient settlements in the Iranian desert akin to the site of Gonabad where there is now no permanent surface water and also lacking conditions suitable to dig a qanat.

As a matter of fact, after the Würm Ice Age, which ended some 10,000 years ago, the central plateau of Iran became warmer and its climate changed to a large extent. The geological evidence shows the existence of lakes into which many rivers poured. The motifs on pottery or inscriptions left from prehistoric time show that there once existed some animals and plants consistent with a much wetter climate in this region. When the environment later changed, becoming more arid, some communities migrated to more favorable regions and some stayed and came to terms with the dry conditions by inventing the technique of qanat. Regardless of the controversial origin of the qanat, the inhabitants of the Iranian desert used qanats in establishing an agricultural system leading to a special economic structure. In this economy there is no trace of what some scholars like Witfogel say about hydraulic civilizations. Karl Witfogel set out a theory on the relationship between water management and government in the east entitled the theory of oriental despotism. According to this theory, in eastern civilizations, from the north of Africa, Iran and India to the Middle Asian steppes, where agriculture depended on irrigation and water was one of the most vital issues, hydraulic despotic governments found the basis to come into existence. This theory claims that the people who were responsible for water management and irrigation gradually formed the germ of the first eastern governments, which later evolved into a dominant aristocracy and then despotic government. In other words, the advocators of this theory believe that the dependence of agriculture on water division systems brought about a totalitarian powerful aristocracy, which paved the way for oriental despotism by suppressing society through economic means, particularly with regard to water management. It seems that this theory is applicable only to regions where surface water streams were available, because in areas whose inhabitants lived off groundwater resources, water was managed by private sector and small owners, so such areas had no potential to house a centralized water management system and accordingly a totalitarian government. That was why the great ancient governments came about in the western part of Iran where big rivers like Aras or Karkhe ran, whereas in the central plateau of Iran, where the qanat system was at issue, the genesis of national widespread governments is unheard of. The reason is that in this area the economic structure is based on water obtained through qanats. This water is limited, and a portion of the qanat's income goes to the qanat itself to maintain it. Therefore, the cultivated lands irrigated by the qanat system could not expand to a large extent and the surplus of the agricultural products was not considerable. In other regions, especially around the big rivers, the surplus of agricultural products could enable the government to equip itself and gain a stronger foothold in society to mobilize people and facilities. But in the territory of qanats the local governments had no access to a strong production system through which they could strengthen their political structure. The people of this region are so peaceable that they never invaded other communities, because the qanat, which was the focal point of their production system was so fragile and vulnerable that they were not prepared to risk it by getting involved in conflict. If the enemy attacked them, the first thing that the invaders used as a military tactic was to cut the veins of the desert. The qanats could easily be destroyed and obstructed just by filling up a shaft well, and immediately the flow water would cease and people might die of thirst, so these people avoided any confrontation with other communities and preferred to manufacture handcrafts to supplements their small income from the qanat rather than invade and plunder other communities, which many tribal people were in habit of doing. Hence, nowadays there is a rich tradition of handcraft in the desert regions whose main water source has been the qanat. Even today, desert towns like Kashan and Naeen are famous for the most precious Persian carpets. The technical characteristics of the qanat have brought about some social, cultural, economic and political traits, which together formed the qanat civilization. Although the above examples are only from Iran, it seems that this situation is applicable to other arid and semi arid regions of the world with qanats.

Here it is worth telling a real story about how these people are interested in their qanats and how they do their best to keep the veins of the desert alive. This vital relationship between people and qanats is rooted in their history and culture. There is a small qanat near Taft in Yazd named Haji Sadegh's qanat, around which a tragic and telling story revolves. Haji Sadegh lived some 60 years ago in a village near Taft and he was a relatively rich landlord. He had several orchards in the village as well as in Taft and other lands irrigated by the same qanat. Once a drought broke out and led to a decrease in the discharge of Haji Sadegh's qanat, which meant so much to him. Haji Sadegh managed to do something to keep the discharge steady, and eventually he made up his mind to extend the end of the qanat's gallery to catch more water. He hired a team of qanat workers to do their best in this respect, but after a short while the qanat master noticed that the tunnel had reached an impasse: a very hard and large rock did not allow the workers to go ahead. The workers refused to continue digging and the qanat master ruled out the possibility of obtaining more water through this qanat, though Haji Sadegh kept insisting on the work. The workers left, and Haji Sadegh had to hire a new team who had accepted to dig through such a hard rock. Haji Sadegh urged the workers to dig through the rock, because he strongly believed in his intuition that the tunnel would eventually open into a good reserve of groundwater. However, the workers lagged behind and could not advance more than a few centimeters a day. Haji Sadegh ran out of money and he had to sell his properties one after another to finance his elusive project. Even his family could not cope with it, and they all left him after he sold his own home to invest in the qanat. He had the qanat dug through the rock hundreds of meters, until he went completely bankrupt and eventually died while observing the workers, but the qanat never came across the water that Haji Sadegh dreamed of. This real story illustrates the fact that in this region the qanat is not only an economic means but a cultural value for which people are prepared to sacrifice themselves.

One of the important legacies of qanat civilization is its water management system, which is extremely accurate in order to enhance the irrigation efficiency as much as possible, and social solidarity is manifest in the qanat water division systems.

3.2 History of Traditional Water Management Systems in Iran

In the course of history throughout Iran, wherever the farm land has needed to be irrigated, there have been water management and division systems. For the major water resources such as rivers, central governments seemingly used to take control of the water management systems, whereas the management of minor resources such as qanats had nothing to do with the state and were largely managed by private individuals, especially the local landlords. The water management systems have always played an important role in Iranian civilizations, and according to some thinkers many economic, political and legal structures in Iran have historical roots in these systems. Throughout a vast area, from northern Africa to Saudi Arabia, Iran, India, and central Asia, agricultural activities have been dependent on irrigation, so watering and water division have underlain the livelihood of the residents of this region. It was impossible to distribute water fairly and avoid conflict over water without a water management system, and some people in charge of water division. Gradually a new social class emerged and took responsibility for water division. On the basis of their authority they gained control over water and established the type of political system that later evolved into the modern eastern ones. According to Witfogel, totalitarianism and despotism could not exist and flourish in western territories where water is so abundant that people did not have to gather around a limited water resource and be submissive to those who controlled it. This is why there was a greater potential in Europe for pluralism and democracy to emerge (Naghib Zadeh, 2000).

The issue of water management has always been a major anxiety for Iranian governments. As an instance, during the Sassanid era, the government had an irrigation department to supervise all water related affairs such as water ownership and distribution. Later, in the 9th century the same department was reconstructed. One of its main missions was to calculate and record the taxes that the owners of water were to pay. Also, when shares of a water resource were sold or bought, this department should have been informed. Experts named *qayyas* or *hassab* were in charge of water division in Abbasid times (Papoli Yazdi and Labbaf Khaneiki, 1998). Taking into account that the political powers were economically dependent on the taxes coming from the agricultural sector, they did their best to regulate this sector, at least whenever they were not in the grip of a war (Bartold, 1970). Water management systems were so important to the state that in most cases the amount of tax was calculated simply based on the amount of available water

(Khosravi Khosrow, 1969). Needless to say, if there were no regulation or legislation on water related affairs, many conflicts and disputes could break out between the farmers, and much time and energy would be needed to settle such problems leading to the destruction of agriculture. So governments tended to intervene in water related affairs in order to prevent such problems, which could eventually inflict damage on the state in addition to the farmers. One good example of the Persian rulers' concern about water management is the ancient dam in Damghan (northern Iran) as Aboodalaf reports. This remarkable dam was constructed by Sassanid kings in order to divide water into 120 shares, each of which belonged to one village. Aboodalaf admits that "he has never seen such a clever and accurate technique to manage water" (Aboodalaf, 1975).

Note that the water management systems were always subject to change. Some factors such as population growth, a change in cropping area, and migration could cause a water management system to change. For example, the immigration of a population of Arabs to the town of Qom, and then a change in their economic background from nomadism to agriculture brought about many conflicts between them and the native residents. After a long dispute over water the natives had to change their water management system so that the Arabs could be included in it. The whole story of how the water management system changed has been detailed in the book "History of Qom" written in the year 957 AD (Qomi, Hassan ib al-Mohammad).

Also, what Maqadasi (945–1000 AD), an Arab geographer, reports on the Merv dam, shows how much the Iranian governments were concerned about the irrigation systems and water management. According to Maqadasi's report, the Merv dam was run by a staff of ten thousand persons who were hired to protect the dam or be in charge of its water management. To measure the amount of water, the dam had a special tablet in an upright position with horizontal lines cut into it. Should the level of dam water reach the sixtieth line, it would imply that the coming year would be so wet and fruitful that the staff of the dam no longer needed to be quite so strict about the water division. But, it could be a bad omen, predicting an upcoming drought, if the level of water did not exceed the sixth line. There were some main outlets in the dam, each of which belonged to a separate village, to distribute water among them as fairly as possible, and then there were other outlets in every village to divide water among the quarters, and more outlets in every quarter, so on. If the dam was short of water, the staff would do their best to decrease all the shares alike (Meftah Elhame, 1992).

The way that water was divided among the farmers was often checked and confirmed by the king or his minister. For example, the water division plan of the Karaj river should be reviewed and then signed only by Amirkabir, the famous minister of Naser al-Din Shah (Enayatollah and others, 1971).

If any problem threw the water division systems into disorder, the government was also in charge of solving it, as King Tahir did about 1000 years ago. A terrible earthquake struck Khorasan province and destroyed many qanats so that their flow completely ceased. After renovation of the qanats, some serious disputes broke out between the owners of the qanats over their water shares and the changes in the place and depth of the qanats. Finally, King Tahir mediated between the owners and settled the problem by calling in all the clergymen and lawyers from all over Khorasan to compile a book on the water division and water related laws (Salimi, 2000). And later, Tahir's actions in terms of qanats turned into the subject of fables: for example, villagers believe that Tahir had a supernatural talent for building qanats, and that he was able to find groundwater sources just by a glance at the earth's surface (Salimi, 2000).

3.3 Water Ownership and Legislation in Iran

There are many rulings on water ownership in Islamic texts. These rulings were the focal point of the Iranian civil laws, which were ratified in 1928. As an instance, Imam Sadiq and Imam Musa bin Ja'far have considered the sale of water, whether in cash or in kind, as a permissible act (Naser Faruqui, Asit Biswas, 2001). According to the constitutional law of the Islamic Republic of Iran article 45, all water bodies and natural streams are common properties to which the government is entitled and which should be used for public profit, and government is responsible for their preservation, issuing permissions and controlling their exploitation. (Ghorbani, 1990)

Also there is some legislation about the ownership and water management of qanats that are of great importance in Iran. For example the law passed on August 28, 1930 takes up the matter of qanats in order to encourage people to develop this hydraulic structure (Ghorbani, 1990). Also on August 28, 1930 and then on September 9, 1930 in terms of qanats some articles were added to this law, which became a basis for the law of water fair distribution enacted on March 7, 1983.

In the year 1942, the parliament passed a law that gave the government more power to regulate water related affairs whether governmental or private. This law let the Ministry of Agriculture set up the Foundation of Irrigation

(Lambton, 1983). Before our water resources were nationalized, anybody could take ownership of a water resource, but in the way in keeping with law. This sort of property was called mobah, meaning something lying within the Islamic territories and not belonging to any particular person. The new legislators adopted the term enacted some laws about water resources which were mobah. According to the civil law article 160, a well or a qanat dug in a mobah land is itself mobah and can be owned by the person who dug that well or qanat (Ghorbani, 1990). But after the enactment of the water nationalization law in 1968, water resources were excluded from the properties considered as mobah to which anybody could hold title under any condition. This law considered all water resources as public property, and according to its first article, the Ministry of Water and Electricity was responsible for the preservation and exploitation of this national property (Safayi, Hossein, 1969). According to articles 23 and 25 of this law, any kind of exploitation of groundwater resources should be only with the permission of the Ministry of Water and Electricity.

Nowadays there are two types of water ownership; the first is ownership of water along with land, and the second is ownership of water independent from land.

A) Ownership of water along with land: this type of ownership is mostly practiced in the regions which enjoy rivers and natural springs. Water and land are bought and sold together, and it is not possible to transact the sale of farm land without the share of water associated with the same land. Therefore the shares of water are usually measured based on the area of the land that is to be irrigated. Note that in qanats that belong to large land owners or landlords, the ownership of water and land are also not separate, and the amount of water varies with the area of land, which is irrigated until nowhere is left dry. This system of ownership is run by the owners themselves and is not as complicated as the ownership of water independent from land (Janeb Allahi).

B) Ownership of water independent from land: this type of ownership is very common in the central plateau of Iran (Lambton, 1983), and mostly practiced in case of small landownership. In this region, what is of greater importance is water not land, since water is not as abundant as land. Regardless of how much land each person is entitled to, the water of a qanat is divided into several shares, which can be purchased by whoever can afford to pay for it. The document of water ownership is separately registered by the notary public office; it details the water shares of the qanat and how many shares the document holder owns out of the all shares of a particular qanat. Also, the specifications and the orientation of the qanat are described in the document. In some cases, a person may possess many shares of qanat water without having any farmland to be irrigated. For example, the qanat of Firooz Abad belongs to 302 shareholders some of whom have no land at all. The separation of water and land paves the way for some individuals from outside the region to acquire some shares of a qanat though they have no land. People who own such shares will usually rent them out to those who own land but not have enough water. In the past, there was a trend in rural regions to invest money in water, which was the most important production factor. According to this type of ownership, water is divided among the shareholders based on minutes and seconds[1], so division of water is a very accurate job needing a special organization headed by a mirab, who is the person in charge of distribution of water among the farmers. Hence,the qanats run by this system of ownership are called mirabi. Research shows that in this ownership system there is no proportion between the area of land and the amount of water a particular farmer may have. In other words, someone may have more water than is in proportion to his small area of farmland and vice versa. This independent water ownership leads to the policy of the unifying of farmlands failing. For example, in the qanat of Qasabe in Gonabad, the ownership of water has nothing to do with that of land, so water sales can be transacted regardless of land (Alvandi, Minoo). In fact, the two production factors which are complementary to each other are bought and sold separately. Needless to say, the economic value of water is not on a par with that of land, and the price of each may independently fluctuate under various conditions. For example, there would be a hike in the price of water in a drought whereas there would be no change in land price. So if someone wants to integrate all the production factors and unify the separate farms to set up agricultural cooperatives, it would be very difficult to specify how much each farmer should contribute and then how much they can share in the profit.

Over the course of Iranian history, water management and agricultural systems have been interwoven, and water management has always been under

[1]As we will describe in more detail, the water of a qanat is used accordance with a rotation, in which everyone's water share comes round at regular intervals. Therefore the division of qanat water is a function of time, and share units are based on time.

the influence of the agricultural systems. In terms of water management, one can refer to two types of agricultural systems as follows:

1. the system of boneh;
2. the system of small land ownership.

3.4 System of Boneh and Water Management

In a nutshell, the boneh is an agricultural unit on which some farmers have the right to work cooperatively. Before the land reform program, the system of boneh was commonly practiced in the central plateau of Iran. According to this system, in a particular village all the farmland belonged to a landlord who divided his lands into agricultural units called boneh, on each of which a team of peasants worked. Each team usually consisted of four peasants, one of whom was the abyar (head of the group and in charge of irrigation), one was the dam abyar (abyar assistant), and two were barzgars (in charge of plowing and cultivating). The abyar had the highest position and supervised the other members of the group, while the barzgar occupied the lowest position in this hierarchy. There was a person called mobasher who acted as an intermediary between the members of the boneh and the landlord.

The water was divided among the agricultural units (bonehs) based on a particular rotation. For example, if a village had 8 agricultural units, each unit had the right to take the water once every 8 days. In other words, the units should have taken turns at being irrigated. But irrigating in this way was not that easy, because in some cases a boneh was not in one location but consisted of different parts in different areas. Therefore the farmers had to distribute their right of 24 hours irrigation throughout the rotation period of 8 days. In some places, because the parts of a boneh were so far away from each other, they would have had to spend the whole 24 hours directing water from here to there rather than irrigating the land. That was why they preferred to irrigate an undivided area, no matter whose lands were being watered, work that was coordinated by the abyars of the bonehs concerned. In the system of boneh, the abyar was familiar with the geographical distribution of land and the way in which the bonehs were irrigated. There was very good social mobility within the hierarchy of a boneh. For example, a barzgar could be promoted to abyar and even mobasher, if he could prove his qualification for a higher position. Also, the abyars, who headed the bonehs, did their best to compete with each other by optimizing the irrigation. Doing so, they could increase their production, which was of great importance to the landlord. The better abyar could manage his team, the more he could produce, and the more he could be promoted in the sight of the landlord. The abyars had to make considerable efforts to satisfy the landlord's expectation, otherwise it was very likely for them to lose their position.

In the system of boneh, the produce was not distributed among the peasants and the landlord very fairly. If the landlord provided the peasants with land, water, plowing ox, and seed, he could appropriate four fifths of the crop for himself, and the peasants had to content themselves with the rest (one fifth). Besides this, other people who gave services to the farmers during a year such as carpenter, mason, blacksmith, and barber also shared in the rest. These are the flaws which made the system of boneh unworkable and paved the way for the land reform program[2], which started in 1963. But this program was not based on a comprehension study, which should have been conducted on the complex relationship between the rural societies, landlords and the government. The land reform program brought no profit to the rural society but turned the peasants into smallholders with no drive to cooperate with each other. If the former regime had properly weighed the advantages and disadvantages of the system of boneh, we probably would not face the problems we do today. Maybe it was better for the government to purchase the lands and substitute the bonehs with cooperatives, instead of granting the lands to individual peasants who could not afford to provide the primary production factors. In this way, the government could have improved on the traditional bonehs by distributing the profit among the farmers more fairly. The government should have watered down the system of Boneh, even though this was likely to provoke widespread protest rather than eliminate a system which carried some advantages. If the system of boneh had been modified, the interaction between the members of the agricultural units would have remained, and in the hope of promotion the farmers would have tried to outdo each other in optimizing water use. This interaction between the farmers in the system of boneh was so efficient that there was sometimes excess water left by which more land could be irrigated, and the farm lands of a village could be extended to 1.5 times (Safinejad Javad, 1989). But the land reform program uprooted the former landlords and made them invest in the industrial and commercial sectors. In the rural regions, whoever had the right to work on the landlord's land could profit from the land reform program by taking ownership of part of that land, and the rest of the villagers had no choice but

[2]Though the land reform law was aimed at supporting farming communities, it could not succeed and yielded the opposite result, because it overlooked the agrarian interactions which had formed over centuries.

to migrate to the principal cities and get involved in only menial jobs. In the rural regions, the water management system lost its traditional function. In the boneh system, only the abyar was capable of irrigating, while the others were experienced in other jobs such as plowing, seeding, etc. Therefore not everyone could irrigate as properly as an abyar after the land reform law was enforced and the landlords' lands were distributed among the farmers. This decreased the irrigation efficiency and caused more water to be wasted. Breaking down the farm lands, and distributing them led to a decline in irrigation efficiency and as a result the search for more water started. Pumped wells were drilled one after another and the many pumps started sucking up the groundwater resources. In the villages, whoever was at a better financial level did not hesitate to drill a deep well, which in fact extracted the water many other people were entitled to. The drilling of pumped wells and the resulting depletion of groundwater, eventually sounded the alarm about the destruction of qanats, and many qanats fell into decay. As a result some farmers fled to the cities, and some had no option but to go to the owners of pumped wells to rent or buy the water they were in need of. There was a boom in the numbers of the pumped wells and more qanats ceased year by year, and the demand for the water mined by the pumped wells increased. In some cases, in dry seasons when there was an urgent need for water, the water became so expensive that it did not make economic sense to keep working on the farms, because the benefit could no longer cover the cost, whereas in the system of boneh, no agricultural unit was short of water, and the benefit and cost were always kept in balance. This system eventually vanished rather than evolved into a system more consistent with the new circumstances of the country. In fact, we wiped out the question rather than find an answer to it (Azkia, 1994). In a nutshell, after the land reform, irrigation efficiency decreased for two reasons:

1. the shareholders were not experienced in the techniques of irrigation except for the abyars who were previously in charge of watering.
2. in the wake of land reform, the competition for a higher position in boneh was over.

In the face of such a decrease in irrigation efficiency, the farmers wanted to keep the level of production steady, so they had to supply more water to their lands in order to offset the shortage of water that it caused. The higher demand for water fueled the problem of over exploitation of groundwater by the pumped wells. This condition made the farmers dependent on the owners of pumped wells, who now played an important role in water crises, especially in the dry season. In addition, the changing of cropping patterns aggravated the problem. Taking into account the changes in people's consumption of water, the farmers tended to produce something which brought them more benefit in comparison to its cost, even though it consumed yet more water and further fueled the water crisis. The only way to put this chaos in order and regulate water consumption in rural regions seems to be traditional water management systems, which deserve to be reviewed. We should consider the water management system as interacting with other socio- economic issues. Needless to say, traditionalism no longer works and in most cases is not practical. Nowadays, the rural societies do not accept the system of boneh with the same traditional shape, and no doubt we fail if we try to rehabilitate it. But it may be practical to mix traditional and modern elements in a way that does not contradict the new conditions ruling our rural societies. To do so, on one hand we should be familiar with the traditional systems and the characteristics inherent in them, on the other hand we should know the new circumstances of the villages, and the influences of the modern world that they are under.

3.5 Small Ownership System and Water Division Management

In some areas in the central plateau of Iran, water has been managed in the system of small ownership that has nothing to do with the system of boneh. One may find two regions whose natural conditions are quite similar but the system of boneh is practiced in one region and the small ownership system is practiced in another. For example, in Sabzevar and Torbat- Heydariye the system of boneh has been common, but there was the small ownership system in Qayen and Birjand, though all these regions enjoy similar conditions. As well as from some mountainous regions whose topography does not allow the system of boneh to take place, there are many regions with the small ownership system.

In the regions now with the small ownership system, such as Birjand, Qayen and Gonabad, there probably existed the system of boneh in pre-Islamic times, which has gradually been replaced with the small ownership system in the course of history. In fact in such regions from the beginning of Islam, there was no possibility for landlords to exist, because the whole area was ruled by local governors who were entitled to almost every property. These governors were the heads of some Arab nomads who had migrated from the Arabian peninsula to this area, and they had a little knowledge of the agricultural production systems, for they used to live off animal husbandry common in their homeland. For example, the governors of Qayen and Birjand came from the tribe of Khozeyme who were forced to migrate from Saudi Arabia to Khorasan by the Caliph Haroon Al-Rashid. Also, Toun and Tabas were ruled by the Zangooyi Arabs who had been brought to Iran by Safavid kings (Yate, Charles Edward, 1986). These local rulers used to charge people a certain tax regardless of the climatic condition such as drought that directly affected the farmers' income. Thus, the local rulers paved the way for the system of boneh to be replaced by the small ownership system, in which the tax was supposed to be paid to the ruler directly, contrary to the system of boneh in which the landlord acted as an intermediary between the peasants and the central government in terms of tax. In this manner, the small ownership system came into existence[3]. This system lacked any abyar who was in charge of irrigation, but an organization for water division emerged. A person called the mirab headed this organization which was responsible for water division, ownership and any transaction of water. A basic concept of this system is the irrigation cycle, which should be defined to better understand the traditional water management of these areas.

3.6 Main Concepts of Traditional Water Management

Iranians traditionally used to live in harmony with their environment, so their techniques to supply water did not end up in the annihilation of groundwater resources. They used qanats as a sustainable technique to extract groundwater, which was recharged in winter and spring directly by infiltration and by special dams constructed by the farmers. To prevent damaging the aquifer, they designated a protected area surrounding the qanat, which was the area defined by a distance between 1 and 3 km from the qanat, depending on the local conditions. The aforementioned dams are no more than a pile of soil upstream of the first and deepest well of qanat (mother well) to catch the floods in winter. The water accumulated behind the dam can gradually penetrate the earth and then seep into aquifer, so an increase in the discharge of qanat as well as the lack of soil erosion are two of the advantages of such dams. Nowadays, most of the dams are leveled and then cultivated with the help of deep pumped wells drilled in the vicinity. The fertile deposits

[3]It is worth noting that in some regions like Yazd, Heart, Mervast, Abarkooh, Ardakan and Meybod, most landlords used to live in the towns and be involved in trading. Such landlords did not create an agricultural system like boneh, but preferred to rent out their lands to individual farmers.

of the dams tempted some farmers to change them into farm land at any cost, even though the qanat would dry up. As an instance, in Yazd a qanat named Chahok-e Nir was recharged by four dams which were located in the boundaries of another village named Pandar. The inhabitants of Pandar had some shares from this qanat, so not only did they put up with the presence of the dams in the middle of their lands, but also they helped the main owners of the qanat with renovating and protecting the four dams. Recently, the farmers of Pandar started selling their shares, and after a while they completely destroyed the dams and drilled some pumped wells in order to cultivate the whole area. The lack of those dams caused the qanat to dwindle. But fortunately such traditional dams provided inspiration for the Yazd Regional Water Authority which is closely concerned with improving groundwater resources in Yazd province. Doing so, they recently implemented some major projects to help recharge aquifers, such as building 18 mud dams being able to inject more than 17 million cubic meters of seasonal flood water into aquifer. This gives hope for the future, where we would be equipped with both tradition and modernity to guarantee a sustainable agricultural system, even though after the land reform program and the advent of modern devices, these traditional water management systems started to fade away.

Another aspect of traditional water management addresses the actions the local farmers take in order to regulate water division, irrigation related subjects and preservation of water resources. They traditionally established some complex systems in order to divide water among the farmers or the shareholders of a water resource. The following topic takes up the subject of irrigation rights based on landownership or time shares within a certain period of rotation.

3.7 Rotation Pattern of Irrigation Water

Water for irrigation is owned by shares. In fact the farmers take turns bringing water to their land. For a particular shareholder, the interval between two irrigations means an irrigation cycle or rotation pattern of irrigation water which may take 6–21 days. For example, if a farmer has an irrigation right of 2 hours within a 6 day irrigation cycle, it means that he has the right to water his land just for 2 hours once every 6 days. The duration of the irrigation cycle ranges from 6 to 21 days in all over Iran, but the average is between 6 and 16 days in most of the country, which is related to the cropping pattern. In terms of wheat and barley which are the most common crops in Iran, the best interval between two irrigations is 12 days, and that is the length of the most common irrigation cycle.

The duration of irrigation cycle varies from area to area with the prevailing cropping pattern, the number of shareholders, climatic condition, etc. For example, the more the number of shareholders, the longer the duration of irrigation cycle. Also, in case of the plants with short and horizontal roots which are more vulnerable to the shortage of water, the irrigation cycle tends to be shorter, because a long interval between the irrigations can do a great damage to the crop. The climatic and soil conditions may affect the duration of the irrigation cycle too, and the fact that porous and light soils have a low capacity to hold water leads to a short irrigation cycle and vice versa. The most essential concept in terms of traditional water management is the irrigation cycle, which specifies when each farmer is to irrigate as well as how many times each shareholder has the right to use the water during a year. Note that an irrigation cycle in a particular region is not always constant, but it may vary for several reasons. For example in the region of Taft (province of Yazd) there is a wide variety in length of the irrigation cycle. Even in a small village, each qanat may have a different irrigation cycle. This variety is attributable to economic, social and climatic factors which are sometimes interwoven. One of the most important causes of the change in the duration of an irrigation cycle is that every qanat needs to be repaired and cleaned once in a while. As mentioned, a qanat is a very long subterranean canal, up to 60 km long, so this system is subject to many natural phenomena which may cause it to flood, collapse, etc. which is why every year the shareholders have to collect a sum of money which goes to the maintenance and rehabilitation of their qanat. In some cases, if the damage inflicted on a qanat is so severe that the shareholders can no longer afford the cost of its repair, they would have to ask a rich person to invest in their qanat, appropriating a day of the irrigation cycle in return. In this case they increase the irrigation cycle, for example from 12 to 13 days, out of which one day belongs to someone who has financed the procedure. In this regard there is a story which shows how such a change in the irrigation cycle may occur. In 1710 an earthquake struck the town of Gonabad and reduced many homes to rubble and destroyed an important qanat named the Qanat of Qasabeh. The obstructions in the qanat were so extensive that the shareholders failed to cope with it, so they requested a rich man named Mirza Ali Naqi Riabi to take part in its repair. He accepted on the condition that two days would be added to the irrigation cycle and given to him in return. At that time the governor of the region was an Arab named Mir Hasan Khan, whose claim for more taxes on the income of the qanat touched off a riot, until Mirza Ali intervened and settled the problem by granting his own shares of the qanat to the governor. The governor was touched by this sacrifice and decided to devote these irrigation shares to the charitable purposes. But after a while, one of Mirza Ali's sons claimed that he had inherited these irrigation shares and now was entitled to them, so a dispute broke out between him and the elderly governor. Mirza Ali's son and the authorities of the great shrine of Imam Reza in Mashhad agreed that half of the shares would be included in the properties of the shrine as endowment, and in return the authorities would back him up until he won his case. From this experience, the shareholders of the qanat have learned that to be content with their own resources is better than to rely upon a stranger.

Sometimes for a religious or charitable purpose, the irrigation cycle may be extended. In many regions, in the name of Imam Hossein who means a lot to Iranian Muslims, the farmers add a day to their irrigation cycle and then rent it out in order to come up with the money they need to hold religious events. This custom is called miyoon or farkhiz, and can only be performed with everybody's consent.

Another reason for changing an irrigation cycle is the fluctuation in the flow of a qanat. If the discharge of a qanat decreases due to drought, the water shares would no longer suffice to irrigate the existing lands. For example, if a farmer has a water share of 4 hours within a 6 day irrigation cycle, and the discharge of the qanat dropped from 100 to 50 liters per second, he would not be able to water all his land. To solve this problem he changes his crop to something more resistant to the dry condition, and then he receives a water share of 8 hours once every 12 days not 6 days. By this means, the existing water can cover all his land, even though the qanat flow is low.

In most parts of Iran, each day of an irrigation cycle has a special name, and usually the days are named after the people who own most water shares. For example, if the first day of the period is named Ali, it means that Ali owns more shares than the other shareholders on the first day. Nevertheless, in the village of Deh-Bala near Yazd, we notice that the names of the irrigation days have nothing to do with any of the shareholders as follows (Table 3.1):

Table 3.1 Names of the days of the irrigation cycle in the village Deh-Bala, Yazd

Day	Name	Day	Name
1	A-Mohseni	7	Haj Mandal Hasani
2	Bagh-e Khajeyi	8	Agha Rasooli
3	Talestani	9	Haji Mohammadi
4	Bagh Anbazi	10	Agha Hadiyi
5	Haj Mandali	11	Jafar Khani
6	Haji Ebrahimi	12	Jafar Khani (Ab-e Kel)

In Deh-Bala, there is a stream named Ab-e Koohi which was once the main source of water in the village. This river was rationed from April to late October, when its water used to dwindle, according to a 12 day period like this: each day of this period was named after a person to whom this water belonged on the same day. Later, when qanats were used more to supplement the shrinking surface waters, the names were retained to mark the days of the irrigation cycle.

3.8 Water Division Units

To measure the irrigation shares the farmers are entitled to, they have invented some units which vary from area to area with the local conditions and the volume of the water available. In sum, there are three types of water division units traditionally used by the Iranian farmers as follows:

1. Units based on the area of farm land,
2. Units based on the volume of flow, and
3. Units based on time.

The first type measures the irrigation share by the area of the field that is to be watered. So if someone has a share of 1000 meters, it means that he is entitled to a volume of water which is able to fully irrigate a 1000 square meter land. This unit is used in regions where there is no scarcity of water. The second type is more similar to the units used today than the others. This type measures the volume of water flow per a particular time.

The third type just measures the time during which a shareholder has the right to irrigate, regardless of either the land area or the volume of flow. In arid areas, units based on time are more common. Harriet Nash has conducted a research project on water allocation at night based on time calculation by the use of stars. According to Nash's study in Oman, which applies mainly to the larger qanat systems, "the water shares are based on the time that water flows to the fields. A twenty-four-hour day is divided into two parts called *bãda* ... related to major divisions of water such as blocks for renting for *falaj* maintenance. It is also divided into forty-eight *athar*. A day-time *athar* is the time it takes for a shadow to lengthen by a man's foot, nominally half an hour. The *athar* may be subdivided but one twenty fourth of an *athar*, a little more than 1 minute, is considered the smallest practicable division for irrigation. Another measurement of time is the *sahm* which is 45 minutes and is used instead of the *athar* in some villages. The sun is still used by many *falaj* communities, even where they have stopped using the stars. The sundial comprises a marker pole and lines oriented approximately north–south, which divide the day into *āthār* or *sahms*. They are relatively easy to use: the shadow of a pole can usually be seen even when it is a bit cloudy.

At night, stars appear to circle the earth and their movement can be used to tell the time for allocation of falaj water. The method of stargazing differs from village to village. In Qaryah Benī Subh stars are watched rising above the horizon. In other villages, walls, palm trees, and posts attached to buildings are used to mark the rising, setting, or zenith of stars. In Qaryah Benī Subh, 21 main stars are used with dividers and other stars, totaling approximately 50. In the Mudaybī area and beyond, twenty four stars are used, but there are minor differences in the 24 stars used in different villages. Stargazing has survived in only a few villages still highly dependent on falaj irrigation and where light pollution is not excessive (Nash Harriet, 2007). A similar method of stargazing has been practiced in central Iran, but there is not much information available about this issue. Harreit Nash, Semsar Yazdi and Majid Labbaf Khaneiki are jointly working on stargazing in Ardakan and Meybod regions to gather the information before being completely forgotten.

The time unit is also very common in the central plateau of Iran. To calculate the time of irrigation, they have invented a special type of water clock or clepsydra (Figure 3.1). Their clepsydra consists of two bowls made of copper, one of which is so small that it can freely float on the surface of water in the large one (Figure 3.2). The floating bowl has a tiny hole at its

Figure 3.1 A water clock for timing irrigation in central Iran.

Source: Yazd Water Museum.

Figure 3.2 A replica of the original bronze bowl found in a bog in County Antrim, Northern Ireland.

Source: www.sciencemuseum.org.uk[4]

bottom through which water can enter the bowl and gradually fill it up. After being filled which may take a certain time, the small bowl sinks in the water and bumps on the bottom of the large bowl. As soon as the bump is be heard, a unit of time is over, the time between the two bumps equals a certain unit of time. One can also find marks cut into the inner side of the small bowl, which divide this unit of time into shorter fragments. The time it may take the small bowl to be filled and sink varies from area to area in the central plateau of Iran. We examined some different types of clepsydra in a number of areas and summarized all the results in the following table (Table 3.2).

Sometimes, in a certain area, the unit of time may vary with the season and the period of rotation within which the irrigation rights have been defined. For instance, in the Bajestan area, the unit of time varies from 2.3 to 17.2 minutes between the months of March and February. In this area, there are three qanats named Mohammad Abad, Golbid and Nowkariz. Bearing in

[4]This bronze bowl is of the type which was employed by the Ancient Britons, probably under the influence of the Druids, for measuring intervals of time. The bowl has a small hole in the bottom, and in use it was placed on the surface of water, which slowly leaked into it until, after a certain interval of time, the bowl sank. The interval was the unit of time; in the case of this bowl, approximately one hour. This bowl bears striking resemblance to that practiced in Iran, despite the fact that the two places are so far apart.

Table 3.2 Duration of time unit in different parts of Iran

Location	Time (Hour: Minute: Second)
Kol-e Birjand	00: 24: 00
Shahik-e Qayen	00: 22: 30
Khor-e Birjand	00: 17: 00
Kadekan	00: 15: 00
Sarbisheh, Zirkooh-e Qayen, Darmian-e Birjand	00: 12: 00
Yazd	00: 11: 15
Zoozan, Boshrooyeh	00: 10: 00
Fakhrabad-e Bajestan, Eshgh Abad-e Tabas	00: 09: 00
Bilond-e Gonabad	00: 08: 30
Gonabad	00: 08: 24
Dihook-e Tabas	00: 08: 00
Khanik-e Gonabad	00: 07: 30
Abiz-e Qayen	00: 07: 00
Aboojafari-e Boshrooye, Kakhk	00: 06: 00
Khosro Jerd-e Sabzevar	00: 05: 00
Serend-e Ferdows	00: 04: 44
Bajestan	00: 04: 36
Tabas	00: 04: 00
Ferdows	00: 03: 00

mind the location of farms and the distance between the qanats and the farms, each farmer may use either one of the three qanats or two/three of them mixed together. The joint flow of the qanats of Golbid and nowkariz only is considered as the standard flow to which all the official proofs refer. For example, if someone claims that he/she possesses ten shares of water, in fact he/she is entitled to irrigating his/her land for 46 minutes, because every unit of time equals 4.6 minutes if the flows of Golbid and nowkariz are together. On the other hand, the rotation of irrigation rights may be based on 21, 14 or 10 days during a year. Within a period of rotation based on 21 days, every shareholder is allowed to irrigate only once every 21 days and so on.

Meanwhile, the length of the period rotation varies from season to season in order to adapt the available water to the existing climatic conditions (Table 3.3). Therefore, there are 15 units of time all of which depend on the period of rotation and the source of water, as you can see in the following matrix. The gray part of this matrix shows 15 possibilities for the unit of time, from 2.3 to 17.2 minutes. Thus, the unit of time would equal 4.6 minutes if the rotation of irrigation rights is based on 14 days as well as both the qanats of Golbid and nowkariz being taken into account.

Table 3.3 Correlation between period of rotation and source of water in the qanats of Golbid and nowkariz

Period of Rotation (Days)	10	14	21	Source of Water
Unit of time (min.)	2.3	3.2	4.8	Mohammad abad + Golbid + nowkariz
	3.2	**4.6**	6.9	Golbid + nowkariz
	8.2	11.5	17.2	Mohammad abad
	8.2	11.5	17.2	Golbid
	5.5	7.8	11.7	nowkariz

3.9 Ownership System and Water Management of Qanats

In most cases, a qanat is run by a private common ownership arrangement which has historical roots in qanat civilization. This ownership system has evolved into a very sophisticated water management system over time. The shareholders of qanats have traditionally established some complex systems in order to better divide water among the farmers or the shareholders, and irrigation rights are based on land ownership or time shares within a certain period of rotation. This water division system is consistent with all likely fluctuations in the volume of water during a year, while quenching the farmers' demand for water. To measure the time every shareholder has for irrigation, they have invented different methods and devices. One of these devices, which has been used in most parts of Iran is a special type of water clock or clepsydra, as mentioned earlier (Figure 3.3).

Figure 3.3 Timing water shares by means of a traditional clepsydra.

Now we know about the irrigation cycle and the unit of irrigation time which are elemental in the water management systems in Iran especially in the Iranian central plateau. In fact, the ownership system of the qanat is associated with these two elements. For instance, if the unit of irrigation time is 7.5 minutes and water is distributed on a 12 day irrigation cycle, this qanat would have 2,304 water shares because: 12 (days) $\times$ 24 (hours) $=$ 288 (hours) $=$ 17,280 (minutes) $\div$ 7.5 $=$ 2,304 (shares). Thus, a shareholder may own one or more water shares, and if we add up all the shares owned by all the beneficiaries, we inevitably reach the number 2,304.

Needless to say, the total shares of different qanats are not equal, because their units of irrigation time are not the same, and the duration of the irrigation cycle varies from qanat to qanat. Hence,the total number of water shares is mentioned together with someone's share to know what percentage of the total shares they own. It is normal to see an ownership document reading, for example, that Mr. x owns 5 shares from qanat Y which has 2,304 shares in total. For example in the qanat of Zarch, Yazd an irrigation cycle comes round once every 15 days, and an irrigation cycle is made up by 140 water shares locally called joreh, a unit of time lasting 10.28 minutes. Thus, the water of this qanat is divided into 12,600 shares during a 15 day irrigation cycle.

In the case of high discharge qanats, the water flow may be divided into separate streams before being utilized. There is usually a special structure named "maqsam" which consists of outlets of different sizes on which the shareholders have agreed (Figure 3.4).

Figure 3.4 A traditional structure named "maqsam" for dividing water into four shares of which one goes to one village and three go to another village (Province of Yazd).

Due to the complexity of the water division among the shareholders, who may number over 500 individuals (as in the qanat of Qasabeh in Gonabad), there are some professionals named mirab who are in charge of the distribution of water among the farms, and are paid a set salary by all the shareholders. While giving water to a shareholder, the mirab also has to consider the time it may take for the qanat flow to get to the given farm. For example, if someone has a right of 46 minutes irrigation, and it takes the water 4 minutes to arrive in his/her farm, then he/she should be allowed to use the water for 50 minutes. Therefore, the mirab does his best to distribute the water among the farms in a way that as little water as possible is wasted along the irrigation ditches. Doing so, the mirab should be familiar with the locations and characteristics of all the ditches leading water to the farms. The mirab has a notebook, including all the irrigation rights in detail, so if the shareholders want to sell or buy any rights they should let the mirab know about any transaction. Unfortunately, nowadays this profession is fading out and nothing is replacing it, so we have witnessed some conflicts over water in rural areas in recent years.

Apart from the issue of qanat water division, the bound of qanat is another important social/technical matter revolving around qanat management. There are some rules in every region to protect the vicinity of qanat and nobody can flout them. If the qanat's depth or length was changed even one meter, the discharge of the adjacent qanats may be affected to a large extent, so mirab or a qanat master should keep track of what the owners of the surrounding qanats may do. The next topic takes up the issue of qanat bound from the social, legal and technical points of view.

3.10 Bound of the Qanat (Harim)

To better preserve the system of qanats, we should be able to define and designate the bound of the qanat (buffer zone) in which nothing harmful to qanat should be done. One of the problems we always face in terms of qanat preservation is trespassing on the bounds or vicinity of the qanat by drilling wells, carrying out the developmental projects, etc. In the past the bound of the qanat was regarded as a very important social and legal issue, but in the wake of the introduction of pumped wells and modern water mining this importance reduced.

3.10.1 Definition of Qanat Bound (Harim)

The term "harim" refers to a special zone we call bound along the qanat tunnel all the way on the surface and some distance from the axis to both sides. This distance ends where the qanat drainage has no impact on water table, varying from qanat to qanat with the hydro-dynamic characteristics of the aquifer through which qanat cuts.

In the qanat bound it is forbidden to drill any kind of well or build another qanat or any kind of facilities which may affect the quality and quantity of the qanat's water. The qanat bound expands as we get closer to the mother well. It is a tradition that the owners of a qanat are entitled to the lands which lie in the bound of their qanat (Aghasi Abdolvahid, Safinejad Javad). We have two separate concepts in terms of a qanat's bound, which should not be mixed up.

1. Bound of the qanat gallery: The bound of a qanat both sides of the tunnel means a particular distance from every point on the tunnel axis to where the water infiltration into the tunnel has almost no impact on water table (Figure 3.5). The area of the qanat bound is at its maximum around the mother well, and tends to be zero near the well between the water transport and production sections. Each qanat enjoys a particular bound (Behnia, Abdol karim).
2. Bound of a shaft well: The bound of a shaft well is an area around the well mouth, which is regarded as restricted to better protect the access well. The diameter of this area is determined by a traditional method named Kolang-andaz. In the past, when a qanat master was asked to determine the bound of a well, he went into the well with his head sticking out of the well, and then threw his pick axe as behind him far as he could. Wherever his pick axe landed, this distance was

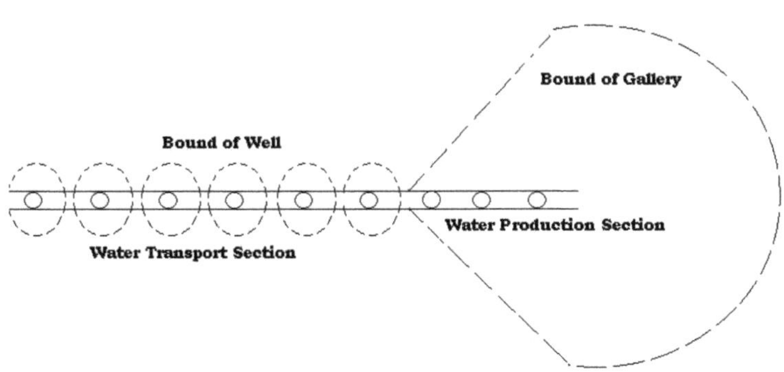

Figure 3.5 Bound of Qanat.

Figure 3.6 Delineating the bound of shaft well by hurling a pickaxe.

considered the limit of the bound area (Figure 3.6). This distance was usually considered to be 3 meters, and the term Kolang-andaz means an area 6 meters in diameter (Aghasi Abdolvahid, Safinejad Javad).

In summary, the qanat bound can be defined both legally and technicaly, though we also see other terms like religious, qualitative and protective bounds which are described below.

3.10.2 Classification of Qanats according to their Bound

To determine the bound of a qanat, it is necessary to pay a visit to the qanat, and the expert can designate the qanat bound based on the type of qanat (plain, mountainous, or semi-mountainous) as well as the type and situation of aquifer.

1. *Mountainous and semi-mountainous qanats*: In such qanats, the groundwater reserve tapped is actually a conduit created by seasonal rainfall. It is even possible that a particular qanat is fed by two separate conduits at two different points. In this case, it is surely harmful to the qanat to drill a well upstream from the qanat, and we should consider this issue while determining the qanat bound. But we can allow another qanat just downstream from the first qanat, because taking water from a downstream conduit has nothing to do with the first qanat's discharge.

The bound of such qanats correlates with the geological conditions and the position of the subterranean conduits and their distance from each other, and in sum their bound is not that large.

2. *Plain qanats*: The bound of plain qanats is larger than that of mountainous qanats. The groundwater reserve on which the plain qanats are dependent is an integrated hydrological unit all over the plain. So extracting water at any point in the plain can affect the discharge of the surrounding qanats. In case of such qanats, drilling a pumped well, whether downstream or upstream, may cause the qanat discharge to dwindle. So we should determine a bound for such qanats, such that the qanats are affected as little as possible.

3.10.3 Legal Aspects of Qanat Bound (Harim)

Harim is an Arabic term meaning "ban". Iranian civil law, article 136 stipulates that it is necessary to define a restricted area around the qanat or river as harim to make the most of them and prevent any kind of damage to them.

If someone managed to re-cultivate barren lands by building a qanat, well or farm, some surrounding lands would be regarded as their bound automatically (Emami Seyyed Hasan). According to article 138 of Iranian civil law, the owner is as entitled to the bound as to the property itself, and trespassing on the bound without the owner's permission is legally forbidden (Emami Seyyed Hasan). Also, according to this article, the bound of qanat and spring in soft soils is 500 meters in diameter and in hard soils is 250 meters. Nevertheless, if these areas are not large enough to prevent damage, the bound can be increased (Emami Seyyed Hasan).

Article 138 addresses just the mother well of the qanat, not all the shaft wells, so if somebody observes this law in terms of the mother well, but builds the water transport section of his qanat near the other qanat, there is no legal restriction therein, unless his action leads to obvious damage to the adjacent qanat, like draining its water (Emami Seyyed Hasan).

In Iran traditionally there are some instructions to determine the bound of a qanat. For example, in the alluvial plains the water production sections of two running qanats should be at least 500 meters apart, and in the arid areas this distance is increased to 1500 meters. In mountainous regions and along the valleys, the bound is considered to be 250 meters at least. Some believe that the bound of qanats in Iran should be between 1500 and 2000 meters in general.

In 1930, the legal gaps in Iranian civil law paved the way for the "Law of Qanats" to be ratified, emphasizing the following issues:

1. Determining the bound of the ownership of water and soil resources
2. Dredging and cleaning the water resources
3. Regulations on the construction of new qanats
4. Regulations for the courts to settle disputes between the owners of qanats and land
5. Regulations on utilization of common qanats (Madanian Gholamreza).

The Law on Qanats passed in 1930 says: "if someone is entitled to a qanat well or a conduit which is located in someone else's land, his ownership over the well or conduit pertains just to the utilization of the same qanat or conduit, and the owner of land can occupy his land outside the bound of the well or conduit and even between the wells and conduits on the condition that his actions do not damage the qanat or the conduit. The bound is the area which has been defined in civil law".

Moreover, one of the most important laws ratified after the Islamic revolution in Iran is the Fair Water Distribution Law. This proposal was put forward to the Iranian parliament and after some modifications eventually it met the approval of parliament in 1982. The topics of this law are as follows:

1. Responsibility of the Ministry of Energy for water
2. Announcement of restriction on temporary exploitation
3. Regulations on construction and maintenance of qanats in barren lands
4. Regulations on rehabilitation of qanats and abandoned wells and exploitation of artesian wells
5. Determining the bound of groundwater resources
6. Water share and permit of rational consumption
7. Maintenance of common irrigational systems
8. Compensating qanat owners for their loss.

The issue of bound has been mentioned in some articles of the Fair Water Distribution Law. For example, article 17 says that the ownership of the qanats, wells and conduits which are located in others' lands is limited to their bound and the owner of the land can use the land around the qanat, well or conduit as long as his activities do not damage those hydraulic structures.

As mentioned, if, in the bound of a particular qanat, another qanat or well is dug, the water of the first qanat would dwindle. So not observing the bound of qanat can bring about a considerable decrease in the discharge of the adjacent qanat and accordingly legal disputes over it. In this case, according to the Law of Fair Water Distribution, the experts of Ministry of Energy are in charge of determining the bound of qanats, wells and water conduits. If a dispute over the bound of qanat breaks out, the relevant courts would process the case after inquiring of the official experts who are authorized by the Ministry of Justice.

3.10.4 Technical Aspects of Qanat Bound (Harim)

There is another concept revolving around the bound of qanats, wells and springs, called hydraulic bound. This bound refers to the area around a water source, in which the water source can be affected hydraulically. In this area no water mining system should be built. This bound is very controversial, and is usually defined by such constant numbers as 500 and 1000 meters, but may vary from region to region. It is very expensive and difficult to determine the technical bound of water sources one by one, which is why this type of bound has remained controversial. To determine the technical bound of qanat and well, we need to do pumping tests and drill observation wells which relatively cost much. This bound, which is also called a hydraulic bound, correlates with such factors as the amount of water extraction in the region, the point of negligible drawdown of the water table, hydrogeologic aquifer coefficients, duration of water exploitation, regional conditions, geological structure and the amount of inflow in the aquifer. There are several methods to calculate the area of effect of well, which show the appropriate area of bound and are also applicable to qanats. One of these is known as the Richards, formula which is an empirical equation first used for qanats by Monzavi.

$$R = 3000\,(h_0 - h) \times k, \tag{3.1}$$

where R is the radius of influence, ie the radius of the bound, h is height of water in the qanat gallery and h0 denotes difference between the heights of the water table and gallery floor in the area before building the qanat, and k is the coefficient of aquifer permeability.

Another is the Dupuit equation which comes from the Dupuit assumption that groundwater moves horizontally in an unconfined aquifer, and that the groundwater discharge is proportional to the saturated aquifer thickness. The formula to calculate the qanat bound is as follows:

$$R = (k/2q)(h_0^2 - h^2). \tag{3.2}$$

R is the diameter of qanat bound (m).
K is the hydraulic conductivity (permeability) of the aquifer that the qanat source zone cuts (m/s).

h_0 is the average height of water table on the edge of qanat bound from the gallery floor (m).
h is the water height in the qanat gallery (m).
q is the qanat discharge (m^3/s)(Behnia, Abdolkarim).

3.10.5 Religious Aspects of Qanat Bound (Harim)

According to religious rulings, a qanat bound is an area 1250 m^2 in diameter in soft soils, and 500 m^2 in hard soils. According to the definition of religious bound, adjacent qanats should run at the same level, otherwise 500 meters should be added to the bound for one meter difference in depth, so that the upper qanats will not be damaged (Fayyaz, Mohammad Reza).

Karaji has said some 1000 years ago that "Abuhanife a well-known Muslim cleric is quoted as saying that if someone digs a well in a place with no owner with the permission of the Imam, he is entitled to that well and the bound of the well should be considered to be 20 meters. If someone builds a qanat, its bound is 250 meters on both sides, and if he digs a well without the Imam's permission his ownership over the well and its bound would be null and void" (Madanian Gholamreza).

3.10.6 Protective Aspects of Qanat Bound (Harim)

The protective bound pertains to the area around the shaft wells, which is necessary to repair and protect the shaft wells whenever needed. This area is some 6 meters in diameter, based on such factors as the position of the qanat, situation of the surrounding lands and their use.

This bound is considered for the qanat gallery too, where the gallery runs shallow. In this case, the bound means a protective area along the tunnel on the ground surface, in which no facilities like buildings or farmlands are allowed. Within this area, which is some 3 meters from both sides, the gallery is susceptible to damage. The protective bound should be such that it would be easy to get access to every point of the qanat, and no activity should be carried out in this vicinity. As shown in the following picture, in Algeria, the qanat wells are marked by round walls that protect them against flash floods and other threats (Figure 3.7).

3.10.7 Qanat Bound from the Water Quality Standpoint

The gigantic growth of cities and development of human societies have led to a dramatic increase in urban sewage, and nowadays billions of cubic meters of

Figure 3.7 Protection of qanat wells in Algeria.

waste water a day is being produced, threatening our groundwater resources. Using chemical fertilizers and pesticides can aggravate the situation by leaching into the aquifer. Industrial waste water can also be serious threat to groundwater reserves. Moreover, the development of mines and washing of minerals as well as leachate from landfills which are not well designed and maintained can contaminate groundwater to a greater or lesser extent.

Therefore, the qualitative bound of qanat pertains to an adjacent area in which potentially polluting activities are not allowed, as well as to the actions we should take to protect the qanats against contamination factors, for example by treating urban, industrial and agricultural effluents before they enter the natural water cycle.

It should be noted that some development programs may trespass on the qanat bound but invisibly. The next part examines the program of land reform which affected many qanats but no rules of qanat bound could restrain it.

3.11 Land Reform: The Shah's White Revolution

The White Revolution was a reform program launched by Muhammad Reza Shah who attempted to import modern economic ideas and government-financed heavy industry projects. The Shah in January 1963 held a national referendum on six measures described under the title of the Shah's White Revolution. In addition to land reform, these measures included profit-sharing for industrial workers in private sector enterprises, nationalization of forests and pasturelands, sale of government factories to finance land reform, amendment of the electoral law to give more representation on supervisory councils to workers and farmers, and establishment of a Literacy Corps to allow young men to satisfy their military service requirement by working as village literacy teachers. Among the package of the White Revolution the redistribution of agricultural lands, which sheared the landed elite of much of its influence, had the most significant effect on the socio-economic circumstances of Iran.

Before the land reform, most of the Iranian population resided in rural regions. Each village consisted of agricultural units named boneh, cultivated for farmers (share-croppers). The duty of each farmer was perfectly specialized. Everybody worked and lived under the management and authority of a landlord who owned the whole village. According to the Law of Land Reform, the villages were purchased from the landlords by the government one after another, and the land was then sold to a few farmers in the same village by installments.

The Land Reform Law was finally implemented, without caring about the majority of the villagers who had no share in the agricultural units (boneh), not profiting from the land reform at all, and without caring about the complicated relationships between the production systems, environment and culture in Iran. The Land Reform Law sped up the destruction of qanats, the bad experience that taught us that development is not a simple concept we can import from the modern world into our own country, without taking our own cultural, economic and ecological conditions into account.

3.11.1 Land Reform and the Destiny of Qanats

Most of the research by the western scholars concludes that the land reform in Iran worked to the peasants' advantage. Lambton (Lambton, 1969) and Warriner (Warriner, 1969) consider the land reform as a social movement that could break down the unfair relation between the landlords and the peasants. Machlachlan (Machlachlan, 1977), Graham (Graham, 1978) and Halliday (Halliday, 1979) believe that this reform was an instinctive reaction of the government to the widespread discontent of the rural communities with government policies. These scholars usually place a high value on the land reform in Iran, as if it were a utopian way through which Iranian society could take a great step toward development and prosperity. But most of them evade answering some important questions such as: what changes did the land reform make to the socio-economic structures of the rural communities? How much land was really distributed among the peasants? Which groups of peasants really profited from the land reform program? etc. We do not deny the fact that some farmers took advantage of the land reform, but we try to weigh up all the long term results and consequences of this program and conclude that it was more detrimental than beneficial.

One can trace the problem to the fact that not all of the peasants gained access to the distributed lands, so the socio-economic structures broke up in disorder and lost their traditional functions. As a result, some appropriate techniques like qanat became separated from their socio-economical context and gradually vanished. To preserve the qanats, the Iranian authorities should have perceived the socio-economic conditions in which qanats could exist, and then they should have simulated such conditions, without necessarily contradicting some required modern programs. On the contrary, the Iranian authorities tried to diminish traditional production systems, while carrying out the land reform program, in order to pave the way for a modern model. They believed that Iran could never achieve a developed stage, unless they completely abandoned the traditional sections of the society that allocated most of the resources to themselves. Therefore, some of the Iranian scholars and politicians tried to exaggerate the technical defects in qanats to justify their own hasty programs and convince farmers to use pump extraction instead of qanats. As an instance, a report entitled "Economic Development of Soil and Water Resources" was prepared in 1966 explaining the amount of water required to irrigate an area of 10,000 square meters or one hectare. The parts of that report associated with modern techniques estimate the amount of water required to be about 10,000 cubic meters a year. But another part, related to qanats and traditional irrigation makes contradictory statements, giving estimates of the amount of water needed of about 16,400 cubic meters a year. There figures represent a thirty percent decrease in the required water in comparison with a realistic estimate when the report explains modern irrigation, and a sixty percent increase when it engages in qanat and traditional irrigation. The report then concludes, from such wrong estimates, that qanats cannot supply the required water to irrigate our farms. Such exaggerating reports resulted in hundreds of qanats being destroyed. As an example, in the area of Yazd alone (in central Iran) there are more than 70 dried up qanats, which have caused many villages and about 2500 hectares of fertile land to be abandoned (Labbaf Khaneiki, 1999). The reason why qanats started drying up is that many deep wells were drilled in lower locations to extract water with pumps. There were also other ways through which the Land Reform Law indirectly affected our qanats, aggravating the seriousness of the situation. One can classify the main impacts of the Land Reform Law on qanats into four categories as follows:

3.11.1.1 Land reform and pump extraction

As we mentioned before, according to the Law of Land Reform, landlords were forced to sell their lands to the government, so that through governmental authorities all the land could be sold to those who worked on the same lands by installments. But mechanized farms were the exception, and having pump extraction was legally considered a proof of it (Azkia, 1994), so most of the landlords were encouraged to replace their qanat with pump extraction in order to save their own lands. Actually, they did not want the government to destroy their traditional position in the rural communities by means of removing their economical roots. They hurried to dig the wells with extraction pumps to avoid being included in the land reform, even if their lands needed no well. Doing so, the number of the deep wells dramatically increased. As an instance, the first well with a pump for abstraction, was drilled in 1958 in Neyshaboor region (in the northeast of Iran). In this region the number of such wells reached 14 in 1960 just when the Land Reform Law was approved and announced, but had increased to 286 in 1970. Massive groundwater extraction causes depletion of finite aquifer reserves, and it dramatically reduces the water table of the whole surrounding area. In the Neyshaboor region, it is estimated that the water table falls about 0.2 meters a year on average, because of the massive groundwater extraction (Velayati, 1999). Therefore, most qanats were drying up, one after another, due to the wells and their pumps, which took the water table away from the access of qanat collection zones (Figure 3.8).

The comparison of qanats with pumped wells can shed light on the fact that such wells are not suitable for Iranian agricultural systems in many cases. Pumping drains the pores in the cone of depression around the well and brings about the soil compaction, sometimes leading to subsidence at the surface, which does much damage to the structure of soil and even to the buildings. If pumping empties karstic holes of water and destroy them, then a circular hollow appears within a radius of 100 meters on surface of earth. But qanat never makes such a problem. The potential loss of fresh water, which makes salt water move toward up stream, is attributable to extractive pumps, whereas qanats never change the quality of water. According to some

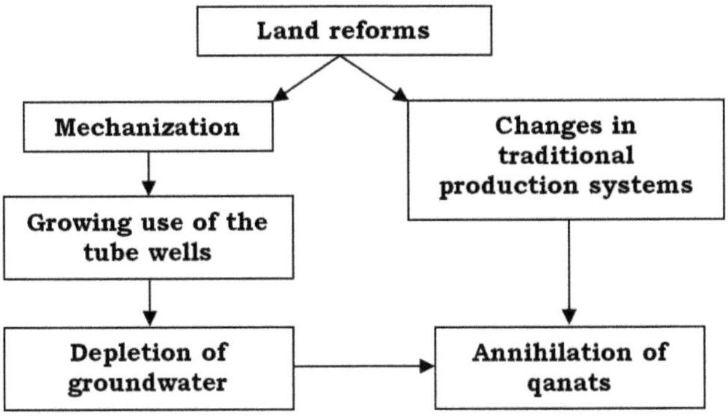

Figure 3.8 Iranian land reform and its impact on qanats.

information related to Iran, the wells with extractive pumps could not last more than 30 years unlike qanats, which last more than 2000 years without any defects. After all, water flows out of qanat only because of the force of gravity, which is free of charge, whereas the extractive pumps consume an enormous amount of fuel per year. For example, in Yazd area there are 4340 wells with pumps, which together consume 205,854,880 liters gas oil a year in order to obtain 926,350,000 cubic meters water. But in the same area there are 2948 qanats, which withdraw 329,870,000 cubic meters water a year without any fuel (Baqeri and Roozbeh, 1999).

3.11.1.2 Land reform and the landlords

Within the agricultural system of Boneh, the lord (Arbab) was usually in charge of qanat. If a qanat would need to be repaired, the lord usually did his best to call in the qanat practitioners (the people who are quite expert at digging qanat) and finance the project. When we were conducting a research on the qanats in Bam region, we noticed that most of the qanat practitioners were not originally from Bam, but they were originally from Yazd who had been hired by the former lords to dig or repair the qanats in Bam. The former lords of Bam did not hesitate to send someone to Yazd region in order to find the best qanat practitioners and bring them along to Bam, because Yazd was famous for the professional qanat practitioners living there. No one except the lords could afford the expenses of traveling to Yazd, hiring the practitioners and bringing them along to Bam.

But the land reform removed the landlord's traditional position, because those agricultural systems carrying such a position were ruined, though anyone or any kind of organization did not exactly replace the lord's role, and on the other hand the surviving landlords did not have enough motivation to take care of the qanats as properly as before. Therefore many qanats were abandoned for a while or even forever.

3.11.1.3 Land reform and the qanat practitioners

In some parts of Iran, qanat practitioners were a professional community having no right to work on the landlord's farms. In the rural regions, society was divided into two casts locally named Nasaqdar and Khoshneshin. Nasaqdar meant the groups who had the right to work for the landlord as farmers on his lands, so they had priority over the second cast (Khoshneshin) who had nothing to do with agricultural activities; however, their jobs would satisfy the other needs of the rural community such as masonry, carpentry, handicrafts and qanat-related activities.

According to the Land Reform Law, the lord's lands should have been distributed just among the Nasaqdars the people who worked on the lord's lands. Therefore the land reform laws did not bring any profits to the qanat practitioners. On the other hand, the qanat practitioners could no longer work for the lord who used to finance the qanat and give them an opportunity to earn a living, so they were encouraged to migrate to the cities or other regions. To our surprise, later in some regions the qanat practitioners themselves became busy drilling the deep wells and sped up the annihilation of the remaining qanats. For example, some of the qanat practitioners moved from the center to the northeast of Iran where there was a great demand for workers being able to drill deep wells in order to install pumps and extract the water. They wanted to make use of their skills to make some money and survive, even though it would result in the destruction of qanats.

3.11.1.4 Specialization of farm workers and land reform

Before the land reform, whoever worked in an agricultural unit (boneh) was responsible for one particular job such as plowing, seeding, irrigating or harvesting, and the farmers rarely interfered with each other's job. Therefore, each farmer could not be as good at all jobs as his own, so most of them were not capable of irrigating the fields efficiently. In a traditional manner, someone who was not expert at irrigation might waste water in many ways, so after the redistribution of the lands the consumption of water increased, and the demand for water quickly surpassed the supply of qanats, mostly due to misusing water. The farmers had to drill some deep well to pump the aquifer to provide the required water, doing so the remaining qanats fell into decay. In an intricate nexus, the aforementioned factors exerted a great influence on qanats, bringing then to the brink of annihilation. Also, uprooting many landlords like Darvish Khan (Figures 3.9 and 3.10) wreaked havoc on the integration of traditional rural society in many ways.

The Land Reform Law was a good example to illustrate that in human society everything is linked. In the developing countries, we have suffered from the lack of a systematic and interdisciplinary view in order to solve our complicated problems. A Persian proverb says that "a wise man never cuts off the branch on which he is sitting", contrary to what some authorities used to do.

Nowadays, environmental and economic disorders resulting from modernization processes have made us think of other ways of production. On the other hand we have been equipped with modern technology while trying to retain the traditional techniques of production. We want to think more of a return to indigenous techniques without ignoring the effectiveness of modern technology. Doing this, we will be able to develop an appropriate agricultural system consistent with the social-economic conditions of Iran.

At present, there are between 34,000 and 35,000 running qanats in Iran, discharging some 8 billion cubic meters water per year without damaging groundwater resources. So Iran has a good potential to prove that if we can perfectly manage the qanats, we can have sustainable production systems

Figure 3.9 Stone garden in Kerman province 45 kilometers south east of Sirjan. This strange garden was built in 1963 in protest at the Shah 's land reform program.

Figure 3.10 Mr. Darvish Khan the owner of the stone garden.[5]

without overusing the existing groundwater resources by means of the modern pumping deep wells. We believe that the Iranians can still benefit from the remaining qanats, perhaps for hundreds of years, as long as they do the following things:

- They take the problem of depletion of aquifers in some basins seriously, and legally extend the restricted areas to stop the over pumping.
- They should refer to the survivors of the experienced qanat practitioners and incorporate their indigenous knowledge into the future programs.
- If they make a decision to change the water management system, they had better learn from the traditional systems, which are in harmony with the environmental and social-economic conditions in much of Iran.
- They should pursue the issue of enhancing public awareness on the qanat and its role in sustainable development.
- They should figure out how to acquaint the new generation with the methods of qanat maintenance and preservation.

[5]Darvish Khan was one the landlords of the province of Kerman. After the Land Reform Law was enforced, the government forced Darvish Khan to hand over his lands to the peasants who worked for him. He was reluctant to do so, but eventually his lands were taken and distributed among the peasants, to his dismay. The only garden which was left untouched was this same garden that later changed to a stone garden. In protest at the land reform, Darvish Khan brought a lot of stones from the hills nearby and hung them from the branches of the trees, as if the trees had borne stone rather than fruit. The landscape of this garden with some 100 dry trees with stones hanging from their branches give a sad feeling to visitors, even though they may not know about the misery of Darvish Khan, who was not even able to talk about his objection because he was born a deaf mute. Darvish Khan never left his stone garden until he died at the age of 83 in silence by his trees in 2007.

3.12 Situation of Qanat Practitioners in Iran

In Iran, the qanat practitioners or muqannis have inherited the accumulated qanat knowledge which is rarely found in other countries with qanats. This fact adds weight to the theory that qanats first originated in Iran, or at least that Iran is one of its oldest origins. Wessels (Wessels Joshka, 2008) who conducted a qanat rehabilitation project in Syria writes in her book "although there is considerable knowledge about the current construction of the specific qanats, we observed that local users have no knowledge of the history and development of the technique of qanat building [in Syria]. There are specific experts such as the *natur, wakil, haqoun* and specialist farmers, but their knowledge is minor compared to, for example, the *muqannis* of Iran. Likewise in Oman, Wilkinson reported that the *awamir* or falaj experts had less knowledge of the surviving methods and specialized constructional techniques than the Iranian experts".

In the past the digging of qanats was one of the most important ways to obtain groundwater in arid and semi arid regions. In central part of Iran the only reliable water source was the water stored in the saturated layers of the earth, and it was only the qanats that made it possible for human settlement. Thus, qanat construction was of a great importance in this area, and accordingly the qanat practitioners, who were the focal point of this profession, could gain a relatively high social position and win society's respect. Because of the high value people used to place on this profession, qanat practitioners were secure in their jobs and relatively well paid, in addition to their high social position. As a result,many people were encouraged to take up this job, though it was awash with many dangers.

Nowadays, the advent of modern technologies applied in groundwater extraction has led the practitioners' experience to be neglected. The experience which has been handed down from generation to generation over tens of centuries is becoming extinct. As a result, this profession is being belittled, and the practitioners' position in society is declining. At present, the young generation has little incentive to take up or stay in this ancient profession, in spite of some modern equipment such as compressor drills, electric elevators, etc. that have made this job much easier and safer.

One of the reasons why the young generation does not show a leaning toward this profession is that a qanat worker is usually subject to a number of dangers, from a sudden cave in to a poisonous snake. It is easy to find stories in Iranian villages about workers buried in a collapse, or choked by the build-up of gas, or washed away by the fury of water. Even today some old workers suffer from pulmonary diseases caused by the conditions inside the qanat.

Another reason is the lack of incentives which can be provided for qanat diggers. Some of the practitioners did not have health insurance or a pension, despite the emphasis put on this matter in the final statement of the International Conference on Qanats held in Yazd, as well as at following conferences, which resulted in a campaign for the authorities to provide for their welfare. This lack of security when too ill or old to work drove many workers to leave the business. Given the dangers and the laborious works which threaten the workers' health, it is necessary to make provision for health insurance and pensions, to encourage them to stay in or take up this profession.

Besides, education and enhancing public awareness about the cultural values of qanats can stimulate motivation and interest among the new generation. In this respect, one of the measures taken has been the setting up of a training center for qanats in which students are being educated with the aid of existing expertise in the desert area of Yazd and the presence of the traditional qanat masters there. This training center, called the Qanat College of Taft was established in 2005. The aim of this training center is to educate young people in qanat technology and educe the chasm between the old qanat practitioners and younger generations. After two years of study these young technicians can handle the maintenance and rehabilitation of qanats. Some 32 students were busy studying there, in April 2010 and 200 young students have graduated from this college so far (March 2010).

Another measure taken is publishing a book entitled "Qanat from Practitioners' Point of View": (Semsar Yazdi, 2004): preliminary considerations indicate that the precious verbal knowledge of the qanat practitioners are in line with modern scientific knowledge about hydraulics as well as to social affairs which are highly worthy of documentation. It is also obvious that acquaintance with qanat construction technology necessitates the interpretation of the local terms concerning the issue. Therefore, ICQHS decided to document the qanat practitioners' invaluable know-how so that the present and the coming generations can have access to a lasting, reliable source of information.

Many factors account for the current unfavorable situation. One of these is low motivation to retain qanats to gain access to groundwater, which nowadays can be extracted by pumped wells more easily.

The Land Reform Law enforced in 1962 also contributed to this problem, because before the land reform many of qanats were landlord-owned, and the landlords could afford the cost of qanat maintenance. Therefore, the qanats were always taken care of, at least for the sake of the landlords' profit. At that time the qanat practitioners were always busy working in qanats all year round. But the land reform made it possible for some peasants to take ownership of the land, although most of them could not afford the cost of qanat maintenance, and some refused to contribute to the qanat costs, unlike other shareholders, so it hindered the usual process of qanat maintenance. Moreover many other of land owners resorted to drilling pumped wells, leading to a scarcity of jobs for the qanat practitioners.

Nonetheless, a considerable number of traditional qanat masters who are over 50 years old are still available and active. They are deeply concerned about the destiny of qanats and are nostalgic for the time when the qanat was the hub of rural life, and they are ready to use all their expertise and experience to save them, though it is no longer of financial advantage to them. We have no accurate statistics of the number of muqannis in Iran, but according to a census taken in Yazd, there are some 2000 qanat masters who are still active. These people were brought up in this same society and it is essential to devise a mechanism by which new muqannis are trained, and this egregious gap between the past and future of qanats is bridged somehow.

References

Aboodalaf, (1975). Aboodalaf's Travel Account. Translated (from ? to ?) by Tabatabyi A. Zavar Publication. Tehran. P. 81–82.

Aghasi Abdolvahid and Safinejad Javad. 2000. Glossary of qanat.

Alvandi, Minoo, Crop Division Based on the Five Production Factors, Articles on the Problems of Peasants and lands, pp. 401–407.

Azkia M. (1994). Sociology of Development and the Lack of Development in Iranian Villages. Ettelaat Publication. Tehran. pp. 85–168/P. 70.

Baqeri M. and Roozbeh M. (1999). Economical Value of Qanat in Comparison with Well. In: International Symposium on Qanat, 293–302.

Bartold, (1970). History of Irrigation in Central Asia (Russian). Moscov, Nauka.

Behnia Abdolkarim. Construction and Maintenance of Qanat.

Emami Seyyed Hasan. Civil Law, Vol. 1, Chapter 3.

Enayatollah R. and others. (1971). Water and Irrigation Techniques in Ancient Iran. Ministry of Energy. Tehran. P. 214.

Fair Water Distribution Law, article 17.

Fayyaz Mohammad Reza, verbal interview.

Graham, R., (1978). The Illusion of Power, London.

Ghorbani, F., (1990), Comprehensive Collection of the Legislations, Ferdowsi Pub., Tehran, p. 15/p. 196.

Halliday F. (1979). Dictatorship and Development, London.

Janeb Allahi, M.S., "Water Division System in Traditional Irrigation in Meybod", Geographical Researches Journal. Vol. 2. Mashhad, P. 57.

Khosravi Khosrow. (1969). Irrigation and the Rural Society in Iran. Social Sciences Journal. Vol. 3. Tehran, P. 49.

Labbaf Khaneiki M. (1999). Postmodern Agriculture: Convergence of Modernization and Indigenous Technology. Geographical Research Journal, 15(58 & 59), 101–118.

Lambton P. K. S. (1969). The Persian Land Reform 1962–1966, Oxford.

Lambton, (1983), Landlord and Peasant in Persia, translated into Persian by Amiri Manoochehr, Scientific & Cultural Publication, Tehran, p. 405/ p. 397.

Machlachlan, L. S. (1977). "The Iranian Economy 1960–1970", twentieth century Iran, London.

Madanian Gholamreza. "Legal Preservation of Qanats and Groundwater resources", Cited from the website of Iranian legislation, www.irbar.com

Meftah Elhame. 1992. Historical Geography of Marghab/Merv/Marry. Historical Researches Journal. Vol. 6&7. Tehran. P. 71–132.

Naghib Zadeh A. (2000). An Introduction to Sociology. Samt Publication. Tehran. P. 64–65.

Naser Faruqui, Asit K. Biswas. (2001). Water Management in Islam, United Nations University Press, P. 105.

Nash Harriet, (2007), "Stargazing in traditional water management: a case study in northern Oman", *Proceedings of the Seminar for Arabian Studies* 37.

Papoli Yazdi M. and Labbaf Khaneiki M. (2003). The Qanats of Taft, Cultural Heritage Organization, Tehran (Iran), P. 18.

Papoli Yazdi M. and Labbaf Khaneiki M. (1998). Division of Water in Traditional Irrigation Systems. Geographical Researches Journal. Vol. 49&50. Mashhad. P. 49.

Qomi, Hassan ib al-Mohammad, History of Qom, translated into Persian by Hassan ib al-Ali Qomi, edited by Jalal al-din Tehrani, pp. 47–49.

Safayi, Hossein, (1969). Civil Laws, Vol. 1, Publication of High Accounting Institute, Tehran, pp. 216–217.

Safinejad Javad. (1989). Traditional Irrigation Systems in Iran. Astan Qods Publication. Vol. 2. Mashhad, P. 244.

Salimi M. S. (2000). The Legend on Creation of Qanat in Shahdad District. The Book of International Conference on Qanat. Yazd Regional Water Authority. Vol. 1. Tehran. P. 158–159.

Semsar Yazdi A. A., (2004). Qanat from Practitioners' Point of View, Iran Water Management Organization, Tehran.

Velayati S. (1999). Critical Factors in Quality Changes in The Aquifer of Neyshaboor Plain. Geographical Research Journal, 15(58 & 59), 119–134.

Warriner D. (1969). Land Reform in Principle and Practice, Oxford.

Wessels Joshka, (2008). To Cooperate or not to Cooperate…? Collective Action for Rehabilitation of Traditional Water Tunnel Systems (qanats) in Syria, Amsterdam Uiversity Press, p. 314.

Yate, Charles Edward, (1986). Khurasan and Sistan, translated into Persian by Rowshani Zaferanloo G., Rahbari M., Yazdan Publication, Tehran, p. 64.

4

A Modern Paradigm

4.1 Definition

One of the disadvantages, that is attributed to the qanat system is the low income obtained through it. The lands irrigated by qanats are usually cultivated in a traditional manner with a low efficiency and, in addition, the cost of this production is relatively high. This cost goes to the maintenance and repair of qanat, fertilizer, seed, preparation of land, irrigation, labor, etc. If we can increase the income that qanat owners receive, their motivation would become stronger to preserve their qanat. This chapter examines the ways in which we can increase this income and accordingly the qanat beneficiaries' motivation to keep their qanats.

4.2 Improving the Qanat's Economic Efficiency

No doubt the qanat system carries both technical and cultural values. From technical point of view, the qanat involves a variety of experimental sciences from mathematics to geology, so it is a feat of engineering that can still play an important role in the rural economy. A qanat is not an object in the museum of history, but it is still crucial in supplying water to cultivated lands in arid and semi arid regions. For example, in Iran this technique discharge over 8 billion cubic meters water a year which is considerable. In some areas, communities still live off this ancient water abstraction system, though it is being threatened by the pumped wells which have mushroomed all over the arid regions. Given that in most cases the ownership of the qanats is private, it seems difficult to set out a comprehensive policy to safeguard this system. For example, in Iran there is some legislation putting a ban on water abstraction by means of pumped wells, because some parts of the country have been designates as restricted areas in which nobody is permitted to pump groundwater. It is to the advantage of the qanats, because if the drawdown of aquifers is halted or reversed, the qanats would have a better chance to survive longer. But the question is how we can persuade the owners of qanats to keep and maintain them, and not to seek alternative water abstraction systems that bring them more profit at lower cost. When we talk to them about the environmental crisis that the over-exploitation of groundwater can bring about in the long term, they still give priority to their own economic situation and the ways in which they can improve their income. If we really want to save the existing qanats, we should seek more practical and feasible solutions than preaching and giving ethical advice to those who benefit from replacing qanats with pumped wells. We should set out a policy so logical that the owners prefer to retain their qanats when they compare them to pumped wells from an economic standpoint. It is true that the cost of drilling and maintaining a pumped well is lower than for qanat repair and maintenance, if case we only use the qanat water for irrigation, but if we add more functions to the qanat system, its economic efficiency will be enhanced, and that is when it makes economic sense in the owners' opinion to retain qanats. In fact we want to figure out how to strengthen qanats to be able to compete with other alternatives like pumped wells.

In this chapter, we mention five possible ways (to enhance the economic efficiency of qanats:

1. incorporating new technologies in qanat maintenance and rehabilitation;
2. breeding fish in the qanat;
3. generating electricity by qanat flow;

4. using qanats in the tourism industry; and
5. setting up some modern irrigational methods.

4.2.1 Incorporating New Technologies in Qanat Maintenance and Rehabilitation

One of the problems that qanats face these days is the shortage of manpower. People are more educated and knowledgeable than their forefathers about the dangers which threaten the qanat workers underground while using traditional tools. In the past the qanat master preferred to hire a boy aged between 8 and 12 years – locally called the lashe kesh – to drag the bucket along the tunnel. The qanat master filled the bucket with the excavated materials and handed it to the boy, who had to drag the bucket along the narrow tunnel to the nearest well, where he hooked the bucket to the rope of the pulley. They hired little boys to do this, because it was difficult for an adult to pass through the tunnel as fast as a little boy. But today it is almost impossible to find a family – at least in Iran – who are prepared to send their sons to do such painstaking labor. They are aware that hiring children for such a job is considered despicable and frowned upon by modern society. Even adult workers are not that willing to risk their life by going down a qanat and using traditional insecure tools sometimes at a depth of 100 meters or even more. Therefore, if we do not find a substitute for the traditional tools, the lack of manpower will put an end to thousands of years of qanat history. In Iran some progress has been made in incorporating modern technologies into the qanat, for example, using an electric lift instead of the traditional windlass, or using an electric lamp or drill (Figure 4.1). Moreover, the idea of using

Figure 4.1 An electric lift which has replaced the traditional windlass.

micro-tunneling technology is catching on in Iran, and maybe in the near future we can replace the human factor with micro-tunneling or even robots as Dr. Kobori suggests (Kobori Iwao, 2005). These measures can enhance the economic efficiency of the qanat, because a qanat system can be constructed and repaired at a lower cost and also much more quickly.

Apart from the application of modern technology to qanats, there are other ways to enhance the qanat's economic efficiency, in particular by increasing the income that owners can obtain from their qanats, refreshing their motivation to better preserve them.

4.2.2 Breeding Fish in the Qanat

Most of qanats are home to a number of species of fish that have adapted to the conditions of the qanat. In Iran, qanat fishes comprised 25 species in Coad's study, representing 40% of faunal species on the plateau of Iran. The number of species per qanat ranges from 1 to 6 although 88% of qanats have only 1–2 species. Qanats in areas with little surface water and low in diversity have only 1 to 2 species, while qanats in better-watered areas with more diversity have 5 to 6 qanat species. The qanat fish are mainly Cyprinidae, which comprise 76% of the total. The Cyprinidae are members of the carp family of freshwater ray-finned fish belonging to the order Cypriniformes. The qanat fauna is a subset of the basin in which the qanat occurs, and as well as fish, includes small species, broadcast spawners, lacking in specialized food requirements (usually scrapers of aufwuchs or feeding on invertebrates), non-migratory, and tolerant of a wide range of environmental conditions (Coad Brian W.). In China, in the Turpan (Turfan) basin, there are three species of fish living in qanats, of which Nemachilus dorsonotatus (current name Triplophysa stoliczkai), also belonging to the order Cypriniformes), is predominant and widely distributed (Lou Ning and Lan Xin, 1990).

In Iran, the local people have no idea where the fish came from originally. In most cases, within living memory there is no evidence that anyone has brought this fish to the qanat, but it is likely that it could have got into the qanat after a water body nearby like a river or lake overflowed, flooding the qanat and introducing the fish. It is also likely that this fish has lived in karstic caves from which it entered the qanat when the qanat cut through the karstic formation. At present, no accurate scientific research has been done on the origin of fish in qanats, but there is some speculation and also some folklore. For example, Anthony Smith says: "we asked the muqannis at the wheel about the fish. They admitted that fish were there and said they lived in every qanat. Birds live in the air and fish live in the water; it was never otherwise. I asked if anyone ever put them there and was told that they came automatically... Everybody denied that anybody had put them there and all agreed that there was no point in doing a thing if nothing was to be gained from it. It is true that no one ever ate them, yet there they were in all except the saltiest qanats... They agreed that a fish's egg was quite large and would have difficulty in seeping into the qanat. They shouted down the shaft for some fish's eggs and up with the next load of mud came a tin full of snails. This story was acclaimed by every Persian, for it was argued that they couldn't be anything else. The little fish within the shell was always pointed out to me. They were equally emphatic that no other animals lived in the channels except bats and snakes, that as both these animals live on air alone and as the fish live entirely on water there was no need for any other form.... On the evening of the first day we were taken on a tour of inspection of Jupar. I was shown the fishes in the qanat and told that on one day a year the largest fish wore a golden crown which it borrows from the treasure that lies at the head of every qanat... If these mountains contain streams that are perpetually flowing and are stocked with fish, then I could understand how the fish colony in a qanat was initiated: for in the spring, when the snow melts and the water cascades down into the plain, great rivers are formed and much land is flooded. Then it would be possible for the fish, or their eggs, to be rushed down with the water and be swept into the well of a qanat" (Smith Anthony. 1953).

In the past, the local people caught these fish for medicinal purposes rather than for food. They believed that if somebody suffering from jaundice swallows a small fish alive, it helps cure the disease. But later, they came to catch the fish for food. The gallery of the qanat, whose length may reach tens of kilometers, provides a suitable place for fish to flourish. In this case, we have two possibilities; the first is to breed the traditional local species in the qanats if the market demands, and the second is to introduce trout, which we know are in demand, to the qanats instead of the local fish. In several qanats in Iran this idea has been put into practice and projects of trout breeding in qanats have yielded very good outcomes. Also in Iran, there are a considerable number of irrigation pools built at the outlet of qanats to build up the head and volume of water. These pools are suitable for fish breeding and the experience in this respect is also promising (Figure 4.2).

At the moment, we are conducting a research project on the optimum conditions for trout breeding in qanats, in order to improve the economic outcome of such projects. Suffice it to say that the combination of water, water velocity, and temperature (between 10°C and 18°C) in many qanats favors trout breeding.

Figure 4.2 Fish in the qanat of Nasr-Abad 70 kilometers from Yazd, Iran.

4.2.3 Generating Electricity

The idea of getting energy from qanat flow is rooted in history, when watermills, the most important qanat related structure built in Iran, were used to grind wheat. Its main parts are a drop tower or water house, two millstones, rotor blades and an axis which connects rotor blades and upper millstone vertically.

The operation of the watermill is based on the potential energy of water due to the depth of drop tower. The deeper the drop tower, the more water's energy is generated. Sometimes the depth of a drop tower reaches 10 meters in order to increase the water pressure.

When a qanat's water reaches the water house, it pours down the well/drop tower and builds up there making a water jet through a nozzle at the bottom of the well. The water pressures, which changes to velocity make the blades rotate, imparting energy to the rotor and to the upper millstone. The lower millstone is motionless. Therefore the friction between the upper and lower millstones turns the wheat into flour. Sometimes, several watermills might be operated by the water of only one qanat. According to the way in which watermills draw on the energy of water, they can be classified into two groups as:

a. potential watermill
b. kinetic watermill

The two types of watermill are based on the difference between the heights of water input and output and water discharge. In kinetic watermills the axis is horizontal, drawing on the water flow directly, whereas in potential watermills a difference between the heights of water inflow and outflow should be built artificially, such that the required energy to run the watermill would be generated. This type of watermill, with a vertical axis is the same as built in the qanat gallery.

As already mentioned, a qanat is a subterranean canal to convey groundwater to the surface, and this canal is intended to surface close to the cultivated area or village where the water is to be used. But in many cases the water table is such that the outlet of qanat is not exactly where the workers want, but some distance upslope from the village. Thus, the workers have to channel this water again by digging a well and an almost horizontal tunnel from the outlet to wherever they need the water. When the water is transferred through such an underground canal, it evaporates less, and is less likely to be taken illegally on its way to the village.

When the water is to be re-channeled in this way, first a shaft well is sunk, from the bottom of which a tunnel is dug, to reach the surface near the village. This shaft well can be the drop tower for an underground watermill, where the build-up of water in the well can provide adequate water pressure to rotate the millstone. At the bottom of the well, a small hole is made such that water can spout out of it and hit the rotor blades of the watermill. Thus, the rotor blades turn and the movement is imparted to the upper millstone by a shaft, which passes through a hole in the lower millstone and then turns the upper one horizontally. There is a gap through which wheat can be poured between the two millstones, whose friction turns the wheat into flour. The distance between the two millstones is adjustable by a handle so that the wheat can be ground fine or coarse. The vertical axis that imparts the spin of the rotor to the upper millstone is braced such that its bottom turns on a piece of iron stone which is fixed in a hole in a tree trunk. The bottom of the axis is sharpened and covered in a metal cap so that the friction between the axis and the iron stone at its base is minimized.

Nowadays this technology has been abandoned: wheat is subsidized and purchased by the government, and is ground in the big factories, so the villagers no longer need the underground watermills to grind wheat. Nevertheless, the idea of generating electricity was inspired by the abandoned watermills, and we can place a turbine in the path of the qanat water, though this time its product is electricity not flour.

According to the classification of the hydropower projects, generating electricity from a qanat is a pico hydro-power project, which includes projects generating between several hundred watts and 5 kilowatts to supply electricity to regions out of the range of the power network for domestic and other limited uses. We conducted a study in the province of Yazd in Iran on the potential that the qanats have to generate electricity. In this area, some 3200 qanats are running, most of which have a head difference less than 8 meters, so it seems that the suitable turbines for these qanats are propeller turbines. Looking at the first diagram below, which gives suitable turbines in relation to the qanat's head and discharge, the turbines of Kaplan and Banki Michell have the highest efficiency, because most of the qanats in this province have a head of less than 8 meters and a discharge below 1 cubic meter per second. This diagram also helps us find out that the limit of electricity generation is less than 10 megawatts. In terms of a low discharge (below 150 liters per second) which is applicable to the province of Yazd, the optimum type of turbine can be determined by the two following diagrams (Figures 4.3 and 4.4). In case we take the minimum required electricity production to be some 400 watts just to provide power needed for light and ventilation in the qanat itself, the qanats with low discharge require a head of 10 meters, which cannot be found in Yazd. But fore qanats with relatively high discharge and low head, the turbines of Powerpal and Nautilus seem suitable.

In closing, we can conclude as follows (1) The maximum electricity extracted from such turbines is 1 kilowatt, but considering the length of qanats which is tens of kilometers it is quite possible to install a series of turbines along the tunnel to get more electricity. (2) Those qanats whose discharge is below 45 liters per second do not meet the requirements of this project in the Yazd area because the generation proportional to the product of net head

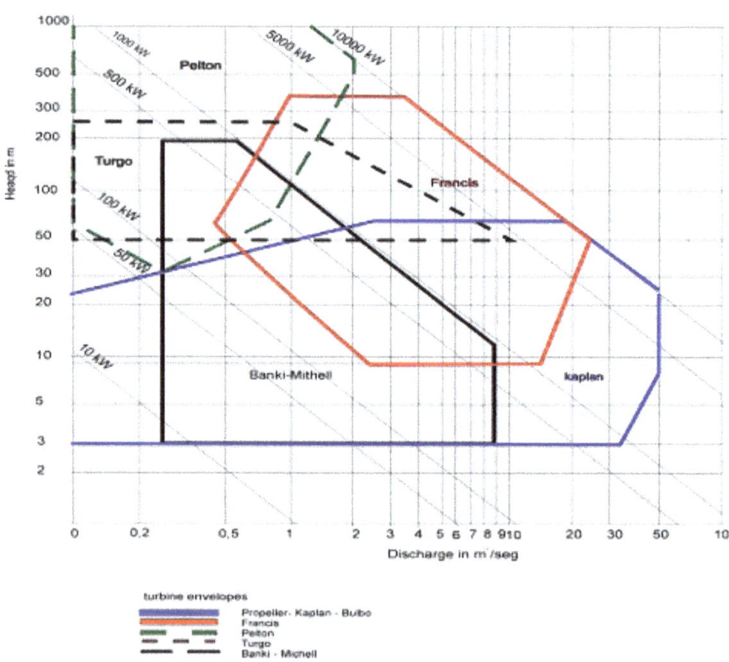

Figure 4.3 Turbines suitable for different ranges of head and discharge.

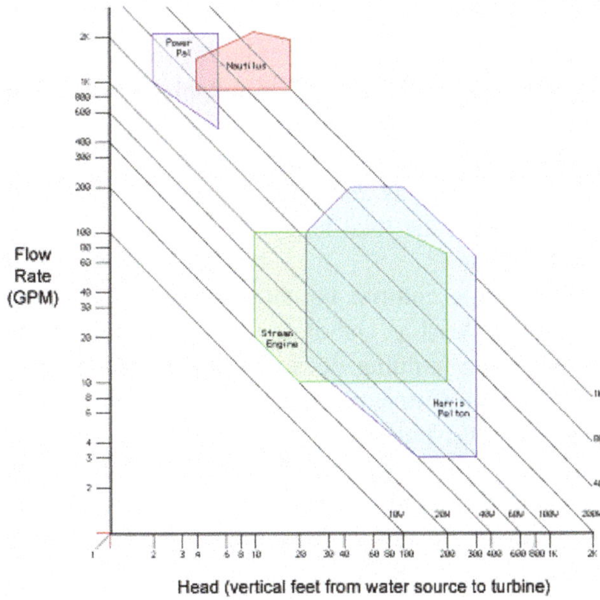

Figure 4.4 Effective operating parameters for various water turbines, showing the relation of head and flow to expected power output of each.

multiplied by discharge, and we would need a higher head than the maximum of 8 meters in the qanats in Yazd.

Due to these requirements, out of 3200 qanats in the province of Yazd, 100 qanats whose discharge is over 45 liters per second and providing an appropriate head have been identified. Each of these 100 qanats can house one or several turbines, such that the total electricity generated by them would amount to thousands of watts (Semsar Yazdi, Labbaf Khaneiki, 2009).

Thus, the qanat, which is an environmental friendly technique and does not bring about any environmental repercussions, is able to merge with another green technology; the hydro turbine, which generates electricity in harmony with nature to satisfy small local demands. This initiative has two advantages:

1. it gives a new function to the ancient qanats and increases their chance of survival.
2. it can improve the economic conditions of the villagers by providing a new income source, taking into account that the Iranian government has a policy to purchase electricity from whoever can generate it through an environment friendly method like wind or water turbines.

4.2.4 Qanats and Tourism

If we add up the galleries of all the qanats running in Xinjiang in China, we would have less than 3000 kilometers of tunnel excavated over time, about which a Chinese geologist says: “these some 3000 kilometers of qanat carry more importance than The Great Wall of China from the viewpoint of civilization”. What might we say about the value of the qanats excavated in Iran, whose galleries reach the total length of some 400,000 kilometers, 8 times longer than the equator?

The qanat is an invisible cultural heritage, running underground. But digging a qanat requires at least as elaborate technology and knowledge as an architecturally interesting building does. If tourists coming to countries with qanats like Iran, Oman, China, etc. know how crucial this system has been in the forming and flourishing of society and civilizations, they would be keen to visit this system and get more information about it. There are many wonders revolving around the qanat from both technical and cultural points of view. For example, the deepest qanat in the world is located in the east of Iran in the province of Khorasan. The mother well of this qanat is some 300 meters deep, whose excavation is a feat of technology given the possibilities that were available thousands of years ago. The longest qanat is some 70–80 kilometers, which can be of interest to those who want to know how people managed to quench the thirst of the desert town of Yazd in central Iran.

With a glance at the engineering methods applied in qanat construction, we realize that many of these methods are unique of their kind. For example, the questions may come to every visitor’s mind of how the qanat masters dig a tunnel underground from one well to another without missing the right direction; or what if the tunnel runs beneath the water table, and how they get the shaft well to the tunnel through the aquifer. The qanat masters have achieved a sophisticated level of knowledge in this regard, which has been handed down from father to son. This know-how is the intangible heritage of qanats that can be attractive to tourists along with the tangible heritage; the physical structure of the qanat and the related structures.

In Xinjiang in Turpan, a high value is placed on the qanat system from the standpoint of tourism. The Karez (qanat) Museum in Turpan is a good example for other qanat holding countries (Figures 4.5–4.7). In fact, the qanat system provides a stable water source all year round in Turpan. They call this system karez and they believe “kar” means “well”, and “ez” means

Figure 4.5 Karez (qanat) museum in Turpan, Xinjiang, China.

Figure 4.6 Karez (qanat) museum in Turpan, Xinjiang, China.

"underground" in Uygur. In 1845, Lin Zexu[1] was banished to the Turpan area. He was deeply impressed by the karez technology and encouraged its spread to other areas. Under his leadership, more than 100 karezes were constructed. Statistics for 1944 show that there were 379 karezes in the Turpan area. By 1952, there were 800, with a total length of 2500 km, equivalent to the length of the Grand Canal. Today there are over 1000 karezes in the Turpan area (Hansen Roger D).

[1]Lin Zexu was a Chinese scholar and official during the Qing dynasty. He is most recognized for his conduct and his constant position on the "high moral ground" in his fight, as a "shepherd" of his people, against the opium trade in Guangzhou. He was replaced by Qishan in September 1840. As punishment for his failures, Lin was demoted and sent to exile in Ili in Xinjiang. While in Xinjiang, Lin was the first Chinese scholar to take note of several aspects of Muslim culture there. (http://en.wikipedia.org/wiki/Lin_Zexu)

Figure 4.7 Karez (qanat) Museum in Turpan, Xinjiang, China.

In 2000, a water museum in Yazd, Iran was set up as a side event during the First International Conference on Qanats (Semsar Yazdi A., 2000). This museum displays the techniques of qanat construction and operation, and its role in the desert towns such as Yazd. Yazd Water Museum is visited by tens of thousands of tourists every year. Apart from setting up water museums in qanat holding regions, some qanats can be prepared for the tourists to visit up close. Thus, the qanat would be turned into a tourist site while keeping its traditional function that is supplying water. In Iran some qanats were customized for this purpose, such as the qanats of Zarch, Ghasabe Gonabad, and Joopar. Attracting the tourist to such qanats serves to enhance public awareness about the qanat system, and can also bring profit to the qanat's beneficiaries and the people living at the site.

Now (2018) qanat tourism has become a booming trend in the Iranian tourism sector. The autours of this book authored another book in 2015 entitled Qanat Tourism which examines the two elements of qanat and tourism as well as the relationship between them. The research area was limited to the city of Yazd, and in the end its findings were put to the test regarding a particular qanat in Yazd named Qasem Abad.

In fact this book was aimed at providing an answer to the question what potentials the qanats really have in the city of Yazd and how these potentials can be deployed in the tourism sector. According to our study, the potentials of qanats which help establish a sustainable local tourism are as follows:

1. Economic importance: although Yazd is not ranked first for the number of its qanats and even for their total discharge in Iran, economic position of qanat and its role in the social fabric of Yazd is more vivid, and

qanat and city are closely interwoven in Yazd from different points of view. Economic importance of qanat makes it easier to preserve this ancient technique. The cultural and economic structures in Yazd are still receptive to qanat, and the chance of its survival would be stronger if we could enhance its economic efficiency through tourism initiatives.

2. Atmosphere of qanat civilization: one of the potentials of qanat in Yazd is the spirit of qanat civilization which can be referred to as qanat lifestyle as well. Qanat civilization consists in a set of cultural, social and economic structures which have germinated and grown on the basis of technical possibilities of qanat, in order to facilitate sustainable interaction between humans and their environment. This environment houses a vast area of arid and semi arid lands where surface streams are extremely rare and people have to subsist on groundwater resources mostly through qanat technique. Therefore qanat as a technology establishes a new relationship between humans and their environment, a relationship that underlies an intricate network of political, social, cultural and economic structures. This same network can be called atmosphere of qanat civilization which is manifest in the adobe architecture, handicrafts, traditional bazaars, orchards, etc. Knowing the role of qanat sheds a different light on many of the cultural phenomena in the city of Yazd. For example orientation of traditional streets and alleys in accordance with the direction of qanat running underneath, the layout of city districts and their correlation with qanat access stairs (Payab), etc all imply the structural and functional importance of qanat in this region. Also handicraft and its complementary role in the qanat economy is another example which can enrich cultural tourism in Yazd. Many of qanats do not extract considerable amount of water in comparison to surface sources, and even this water dwindles or dries up at all during a drought. People of Yazd used to adopt some wise strategies to stave off the economic consequences of such droughts and one of them was handicraft in order to compensate for the income deficiency of qanat based agriculture. Handicraft evolved and developed in that same atmosphere of qanat civilization. Our knowledge on such a relationship between different elements of qanat cultural landscape leads to a rational, purposeful, informative and enjoyable tourism.
3. Systematic relationship between qanat and other historical and cultural attractions: another potential that qanat has is its either direct or indirect relevance to other historical and cultural attractions in Yazd. Qanats are not just some technical objects aloof and independent from urban

life. In contrast, qanats are like a network of veins spreading all over the body of city to bring life and prosperity everywhere. Therefore it is very normal to come across tens of beautiful payabs, adobe houses, watermills, cisterns, mosques, fire temples, etc if we follow the direction of a qanat in the city. The existence of such attractions in the vicinity of qanat can enliven tourism sector in Yazd by inviting more tourists who are in search of the secret of prosperity in desert. A visit to a historical qanat is akin to seeing tens of historical relics related to qanat, and it would be a journey down deep to the heart of ancient Yazd.

4. Easy access: relatively easy access to qanats in Yazd is another potential to develop such tourism in there. In the past there were over 70 qanats running beneath the city, and now at least 8 qanats are still running in the present urban vicinity, including the qanats Qasem Abad, Qasem Naqi, Rahmat Abad, Hassan Abad, Zarch, Najaf Abad, Kheyr Abad, and Shehneh. It is possible and easy to get access to the exit of these qanats and pay a visit to their galleries through their payabs in Yazd.
5. Structural and physical values: qanats of Yazd originate from the mountain range of Shirkooh in south and southeast and travel tens of kilometers to the city. Considerable length of these qanats which passes through a variety of geological formations provides a unique opportunity for those interested in geo-tourism to enjoy.
6. Historical values: Yazd has always been indebted to qanats for its prosperity since its dawn. Each qanat of Yazd denotes a particular historical event or crucial dignity whose stories make tourism in the city more attractive.
7. Intangible cultural values: the tradition of qanat construction and maintenance is deeply rooted in history and has given rise to a precious set of know-how, expertise, customs and jargons all revolving around qanat. These intangible cultural values can be an important part of qanat tourism (Labbaf Khaneiki, Semsar Yazdi, 2015).

4.2.5 Modern Irrigation Methods

One of the measures that can enhance the economic efficiency of the qanat is to incorporate new irrigational methods like trickle irrigation. Thus, we can irrigate a larger area with a particular amount of water, resulting in a higher economic efficiency. The qanat of Mehdi Abad in Yazd in central Iran is a good example of using qanat water in a more modern way. In the past this qanat, which originates in Mehriz, surfaced in Yazd to irrigate some farms in the outskirts. Later, as the city developed, the outlet of the qanat was encompassed by new homes and municipal facilities and the lands which were once cultivated were turned into a part of the city. The water flowing from the qanat remained unused. A charitable foundation managed to purchase the qanat from its former owners, who could no longer use its water and had no idea how to make the most of this water which was the most precious thing in such an arid region. After they took ownership of the qanat, the charity found another place upslope from its old outlet, whose topographical condition allowed the water to reach the earth surface. This area was such that they could set up a large orchard irrigated by the water flowing out of the new outlet of the qanat. They built a large pool, coupled with a state of the art irrigation facility to optimize irrigation efficiency. Thus, they could plant some 120 hectares of land with pomegranates, by means of a qanat whose discharge does not exceed 50 liters per second. Another way to promote water efficiency in qanat use is to construct an irrigation pool. Many qanats have a low discharge rate of just a few liters per second, which cannot fulfill the farmers' needs. If we build up this water in a pool, we can increase the flow rate, admittedly over a shorter period, by 5–10 percent. Now this discharge is able to reach further distances and irrigate larger areas. Therefore, another option to enhance the efficiency of qanats is to build such irrigation pools where the topography is suitable. Since two years ago this method was introduced in many qanats in several countries and the farmers continue to use it.

References

Coad Brian W. (www.briancoad.com)

Hansen Roger D., "Karez (Qanats) of Turpan, China", at http://www.waterhistory.org/histories/turpan

http://en.wikipedia.org/wiki/Lin_Zexu

Kobori Iwao. (2005). A History of Water Issues, United Nations University, New York, p. 190.

Labbaf Khaneiki Majid, Semsar Yazdi Ali Asghar. (2015). Qanat Tourism, Yazd: International Center on Qanats and Historic Hydraulic Structures (UNESCO-ICQHS).

Lou Ning and Lan Xin. (1990). The distribution pattern of vertebrates with particular relation to karez in Turpan basin, International Symposium on Karez Irrigation in Arid regions, China.

Semsar Yazdi A. (2000). A Report on the 1st International Symposium on Qanat.

Semsar Yazdi A. and Labbaf Khaneiki M. (2009). Extracting Electricity from Groundwater Flow; A New Environment Friendly Source of Energy, International Conference on Environment and Electrical Engineering, co-sponsored by IEEE Poland, Karpacz, Poland.

Smith Anthony. (1953). Blind White Fish in Persia, E. P. Dutton and Co., Inc.

5

Other Countries

5.1 Present Situation of Qanats in the World

Qanats have played a vital role in groundwater extraction since very ancient times. They were the focal point of the formation of civilization in various parts of the world. The communities whose main source of water had been qanats, made their utmost endeavors concerning promotion of exploitation and management of such an amazing technology.

Since the introduction of pumped wells, many of the qanats have been phased out. But in some countries such as Iran, Oman, Afghanistan, Pakistan, China, Azerbaijan, etc. some qanats are still operational and supply a considerable amount of water to the agricultural sector.

Henry Goblot who wrote the book "Qanat, a technique of water acquisition" reports that at the time of his book's publication, the qanat system was in active use in Iran, Afghanistan, Pakistan (Balochistan), China (Turfan), Iraq (Erbil), Syria (Aleppo), Saudi Arabia, Morocco (Morocco city and Tafilat area), Algeria (Tablbala oasis), as well as in Egypt, Libya, North Algeria, Spain, and Spanish America where just some relicts of this technique have been found and their qanats were no longer active (Goblot Henry, 1979).

Goblot omitted the qanats of Oman and Jordan which were active whether at that time or at present. Anyway, at present (2010) according to existing reports, some countries, e.g. Iran, Afghanistan, Pakistan, Oman and China enjoy active qanats. Reports from Azerbaijan (the enclave of Nakhchivan and the mainland), Iraq, Syria, Saudi Arabia, Morocco and Algeria show that a small number of qanats are running in those countries. It is quite possible that there are running qanats in some other countries but we are not aware of them, and hence, we do not rule out the existence of the qanat system in countries which are not mentioned here.

Some accurate information in terms of the situation of qanats in Iran, Afghanistan, Pakistan, and Oman has so far been received, as follows.

5.2 Iran

In Iran, the system of qanats has played a vital role. Historical records confirm that some qanats in Iran are very ancient, dating back to before Islam. More information about the antiquity of Iranian qanats can be found in chapter two of this book.

The longest qanat is the qanat of Zarch with a length of 80 km and the deepest is the qanat of Gonabad that is about 300 meters deep. The exact number of qanats in Iran is controversial. In the years 1984–1985 the Ministry of Energy took a census of 28,038 qanats whose total discharge was 9 billion cubic meters. In the years 1992–1993, the census of 28,054 qanats showed a total discharge of 10 billion cubic meters. Ten years later in 2002–2003, the number of the qanats was reported to be 33,691 with a total discharge of 8 billion cubic meters that irrigate 14 percent of irrigated farm lands.

According to the last report, released in 2018 by Iran Water Resources Management Company affiliated to the Iranian Ministry of Energy, there are 41,031 qanats with an annual discharge of 4531 million cubic meters in the country. This amount is 7.3 percent of all the groundwater discharged through deep pumped wells, semi-deep wells, qanats and springs.

5.3 China

The origin and diffusion of the karez (kahriz) or qanat system in China is still controversial among the historians. Altogether there are three speculative ideas about this issue; it is likely that this technique originated from inland China (Shensi); it is possible that Persian influence probably around the 18th century brought the karez to China; or this technique might have been invented by Uighur people independently (Shouchun Wang, 1990). Chi-fei believes that the karez in Xinjiang was invented and developed by Xinjiang local people who struggled with aridity for a long time. He quotes a Russian geologist as saying "Xinjiang karez has some special characteristics which are different from the Central Asia and Iran karez in construction techniques and utilization" (Chi-fei Liu, 1990). Apart from these principle theories on the origin of karez in China, Zhang Xiru uses "materialism dialectics" to extract a new concept on their origin. He considers a multiple origin: "The karez in Xinjiang was progressively established by the people of all nationalities in Xinjiang through their practical work on agricultural production and struggle with drought and disasters" (Xiru Zhang, 1990).

In fact, most of the karezes or qanats are/were concentrated in the region of Turpan in the deserts of northwestern China in Xinjiang Uighur Autonomous Region (Figures 5.1 and 5.2). Turpan has long been the center of a fertile region and an important trade center along the Silk Road's northern route, at which time it was adjacent to the kingdoms of Korla and Karashahr

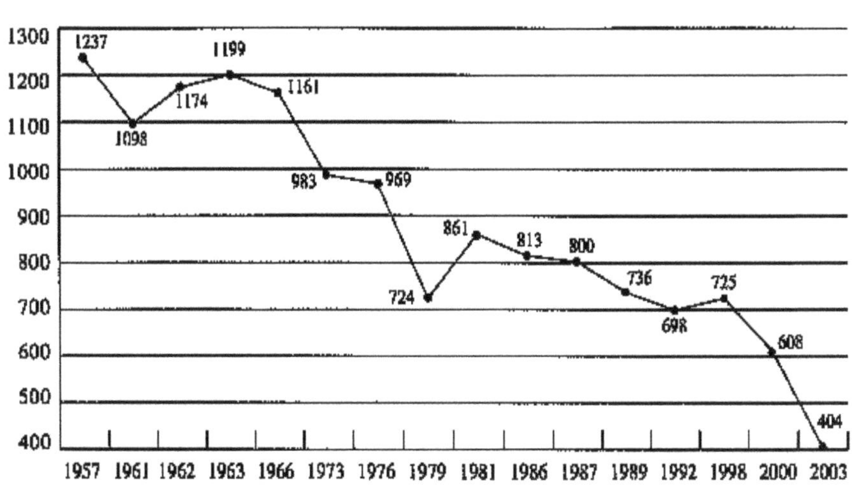

Figure 5.1 Diagram of the number of Karez (qanat) in Turpan region.

Source: Kobori Iwao, 2009.

Figure 5.2 Workers repairing a karez (qanat) in Turpan on June 1, 2005[1].

to the southwest. The historical record of the karez system extends back to the Han Dynasty (beginning in 206 AD). The number of karez systems in the area is slightly below 1000 and the total length of the canals is about 5,000 kilometers in length (Roger Hansen). According to Kobori, in 1957 the number of karezes in Turpan amounted to 1,237 and this number dropped to 404 in 2003 (Kobori Iwao, 2009). Despite the fact that the number of karezes has fluctuated time, the trend is now downwards, at least since 1957.

5.4 Tunisia

The foggara water management system in Tunisia, used to create oases, is similar to that of the Iranian qanat. The foggara is dug into the foothills of a fairly steep mountain range such as the eastern ranges of the Atlas Mountains. Rainfall in the mountains enters the aquifer and moves toward the Saharan region to the south. The foggara, 1 to 3 km in length, penetrates the aquifer and collects water.[2]

Unfortunately we did not have access to more information about the qanats in Tunisia at the time of writing this book. We hope to receive more

[1]The project to save and protect karez in Turpan, which plans to repair and reinforce 440 karezes in three years, has passed experts' appraisals and been submitted for approval. The project will hopefully stop the sharp decrease of karez numbers. english.people.com.cn/200506/02/archive.html

[2]"Water: symbolism and culture" http://www.institut.veolia.org/en/cahiers/water-symbolism/water-symbolism/practical-issues.aspx

detailed information about the history of qanats and their number and total discharge by the next print of this book.

5.5 Algeria

Qanats (designated foggaras in Algeria) are the source of water for irrigation at large oases[3] like that at Gourara. The foggaras are also found at Touat (an area of Adrar 200 km from Gourara). The total length of the foggaras in this region is estimated to be thousands of kilometers. Although sources suggest that the foggaras may have been in use as early as 200 AD, they were clearly in use by the 11th century after the Arabs took possession of the oases in the 10th century and introduced the Islamic religion. The water is metered to the various users through the use of distribution weirs which meter flow to the various canals, each for a separate user. The humidity of the oases is also used to supplement the water supply to the foggara, as illustrated below (Figure 5.3).

The temperature gradient in the vertical shafts causes air to rise by natural convection, causing a draft to enter the foggara. The moist air of the agricultural area is drawn into the foggara in the opposite direction to the water run-off. In the foggara water condenses on the tunnel walls and the air passes out of the vertical shafts. This condensed moisture is available for reuse.

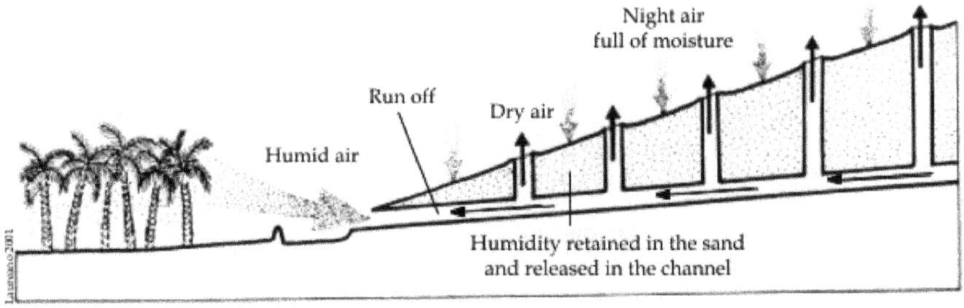

Figure 5.3 Humidity water supply to the foggara.

Source: www.mappeonline.com

[3]An oasis is an isolated area of vegetation in a desert, typically surrounding a spring or similar water source.

The oasis is an ecological system whose initial supply of condensation and moisture is extended by planting palm trees that produce shadows and attract organisms, thus forming humus (Laureano Pietro).

Cirella writes that in Algerian oasis, temperature ranges from night to day thus causing overnight condensation on the soil. If hidden precipitation is carefully managed, important water reserves can be harvested. Proper hydraulic devices enable to harvest water vapor in the atmosphere, thus keeping it in the subsoil before it evaporates during the day. This is the way some typical foggara networks of Touat[4] are fed. Since they are not dug deeply in the soil they are called surface foggaras (Cirella Anna). Also, in the UNEP Global Desert Outlook, it is mentioned that condensation of atmospheric water vapor is the third source of water supply in the foggara system (Laureano Pietro and Sciortino Maurizio, 2006). According to personal correspondence with Mr. Taha Ansari, the National Agency of Hydraulic Resources (ANRH) that manages water resources in Algeria has announced the number of foggaras in the region of Adrar to be 1,416 out of which 855 are still used to irrigate palm gardens. In this region, 6,218 pumped wells have been drilled in the vicinity of foggaras and As a result,many of them have dried up (Figures 5.4–5.6).

Figure 5.4 Foggara in Algeria.

[4]Oasis group, west-central Algeria. Situated along the Wadi Messaoud (called Wadi Saoura farther north), the Touat oases are strung beadlike in a northwest-southeast orientation west of the Plateau of Tademaït. The area was brought under Islamic control in the 10th century AD. In modern times the mixed population of Arabs, Berbers (Imazighen), and Ḥarāṭīn (dark-skinned agricultural workers) effectively resisted French subjugation until the early 1900s. The area passed to independent Algeria when the French surrendered control in 1962 (www.britannica.com).

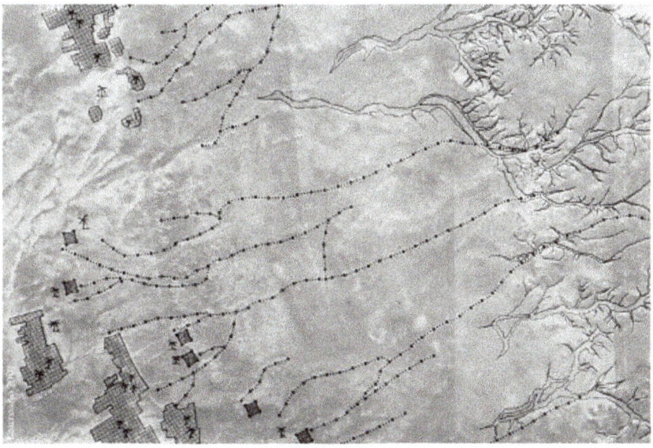

Figure 5.5 Aerial view of part of the foggara network supplying the oases of the Sebkha of Timimoun (Algeria).

Source: www.mappeonline.com

Figure 5.6 A traditional device for measuring the flow of qanat water in Algeria.

5.6 Morocco

In Morroco, on the margins of the Sahara Desert, lies the isolated oasis of Tafilaft relying on qanat (locally called khettara) water for irrigation since the late-14th century (Figures 5.7 and 5.8). Qanats are also located in the Haouz

Figure 5.7 Khettara in Morocco.

plain, around the city of Marrakesh and to the south of the Tensift River. Here they are fed by aquifers recharged by runoff from the High Atlas Mountains close to the mountains. The aquifers in this region are of porous limestone and, in the plain areas, they are also subject to recharge from irrigation water and from rainfall. The average length of khettara in the Haouz has been calculated to be 4 kilometers and the average discharge to be 10 liters per second. In the 1980s (?), around three quarters of the khettaras produced a discharge of less than 10 liters per second and 93 percent produced less than 20 liters per second (Joffe, 1989).

Up to 600 khettaras have been identified in the Marrakesh region alone and in the 1960s 13 percent of the irrigated territory in the Haouz was irrigated solely by khettaras and a further 7 percent by khettaras coupled with other techniques, usually wells, although khettaras were 10 times as important in generating irrigation water (Joffe, 1989).

In Marrakesh and the Haouz plain the qanats have been gradually abandoned since the early 1970s as they have dried up, because the catchments of the tunnels were overexploited through wells and pumps that dried up the aquifers (Ruf Thierry, 2008).

In the Tafilaft area, half of the 400 khettaras are still in use. It is deemed that some developmental projects such as the Hassan Adahkil Dam have also

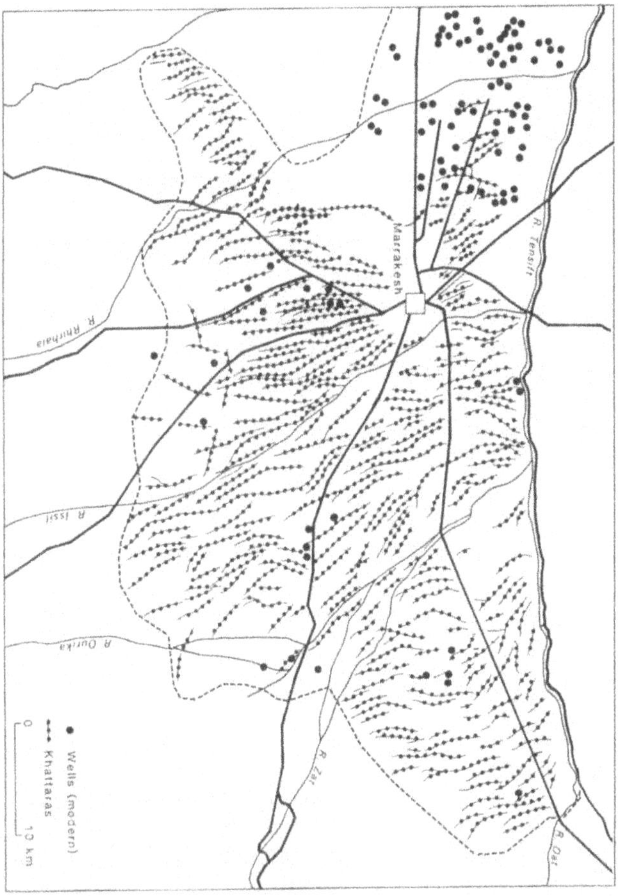

Figure 5.8 Locations of khettaras in the Marrakesh irrigation region.

Source: Cathary M., 1964.

had a negative impact on local water tables, leading to the loss of half of the khettara.[5]

Joffe believes that the main reason for the abandonment of khettaras is the very high labor input required to maintain them – it takes three laborers two days to repair each shaft well – together with the increasingly easy access to modern diesel or electric pumps (Joffe, 1989). The haritin were the hereditary class of qanat diggers in Morocco who used to build and repair these systems (Nair Sankaran, 2004).

---

[5]http://www.waterhistory.org/histories/qanats/History from [Waterhistory.org]

5.7 Egypt

All that is mentioned in the literature about the qanats in Egypt is related to the past and as far as we know, this country has no active qanats. Goblot writes in his book that the common opinion is that the technique of qanats was brought to Egypt by the Persian conquerors for the first time in the reign of Darius I (522–486 BC). In fact the qanats in Egypt were built during two separate periods; some were built by the Persians at the time of their presence, and some were built by Romans during their long period of occupation, from 30 BC to 395 AD. In any event, the magnificent temple built at the time of Darius shows a time of wealth and prosperity. Such activities as building qanats brought prosperity to this oasis and brought together and nourished a population of between 8 and 10 thousands, with 60 thousand palm trees and such crops as barley, wheat and rice (Goblot, 1979).

There are 4 main oases in the Egyptian desert, and the Kharagha Oasis has been extensively studied. As early as the second half of the 5th century BC there is evidence that water was being used via qanats. The qanat was excavated through water-bearing sandstone which seeps into the channel to collect in a basin behind a small dam at the end. The width is approximately 60 cm, but the height ranges from 5 to 9 meters; it is likely that the qanat was deepened to enhance seepage when the water table dropped (as is also seen in Iran). From there the water was used to irrigate fields (Wuttmann Michel, 2001).

There is another instructive structure located at the Kharagha Oasis. A well which apparently dried up was improved by driving a side shaft through the easily penetrated sandstone (presumably in the direction of greatest water seepage) into the hill of Ayn-Manâwîr to allow collection of additional water. After this side shaft had been extended, another vertical shaft was driven to intersect the side shaft. Side chambers were built and holes bored into the rock — presumably at points where water seeped from the rocks — are evident (Wuttmann Michel, 2001).

5.8 Spain

Spain enjoys a long history in the utilization of qanat systems. Turrillas in Andalusia on the north facing slopes of the Sierra de Alhamilla has evidence of a qanat system. Granada is another site with an extensive qanat system. Goblot believes that the qanats of Spain have something to do with the Arabian-Persian influence on the Iberian peninsula and he takes this influence for granted (Goblot, 1979). Qanats can be traced in many parts of Spain, for example in Madrid where it is called "viajes de agua". Gil Clemente reports that the only source supplying water to Madrid was the qanat system during the 17th, 18th and early 19th centuries. The total length of these qanats amounted to 124 kilometers out of which 70 kilometers was the water production section and the rest was the water transport section (Gil Clemente, 1911). Their total discharge was reported to be 3,600 cubic meters per day, but this dwindled over time, dropping to 2,000 cubic meters by the mid-19th century (Canal de Isabel II, 1954). One of the qanats of Madrid which has been renovated and now is operational is that of "la Fuente del Berro" which has two branches. This qanat that had long served Madrid was phased out in 1977 because of organic contamination of its water, but later was rehabilitated, cleaned and made operational again in 1983 (Maria Bascones Alvira, 2001). There is one other qanat in Madrid called "San Isidro", but the other qanats no longer flow (Bernardo López Camacho, 2001). Another important qanat of Spain still running is in the town of Ocaña. This qanat is a feat of engineering, which is probably of Roman origin though the trace of the time of Arabs is seen in it as well. In 1976, this qanat was registered in the national heritage list of Spain. It is likely that the name of the town Ocaña has something to do with its qanat; Canna – Caño (tuyau) – qanat (Bernardo López Camacho, 2001). In summary, the qanat has a long history in Spain and has been used as a water supply for centuries (Figure 5.9). Goblot attributes

Figure 5.9 The third access well of a 650 meter long qanat in Pozuelo de Alarcon, Spain.

Source: www.geologossinfronteras.org

the qanats of the new continent to the Spanish explorers who set foot in there for the first time. According to him, some of the explorers who were familiar with the techniques, managed to build qanats to obtain groundwater where the surface sources were not enough. Thus, this technique turned up in America, spreading to California, Mexico, Peru, and Chile.

5.9 Italy

The entire ancient town of Palermo has been built over a huge qanat system built during the Arab period (827–1072 AD). Many of the qanats are now mapped and some can be visited. An interesting building is the famous Sirocco room, which has an air refreshing system using the flux of waters of a qanat and a wind tower, a structure able to catch the wind and direct it into the room.

Pietro Todaro reports on the qanats in the plain of Palermo, which are locally called "ingruttato". He believes that the Islamic influence in that region paved the way for the qanat system to be developed (Todaro Pietro, 2000).

Also, in the district of Naples, the Roccarainola qanat locally known as "Acquedottodelle Fontanelle" (Small Fountains' Aqueduct) is still running. It has been very important for local communities for a long time, and was been the only drinking water source until the 1950s, when a modern aqueduct was constructed. The qanat has a small fountain of water located near its entrance as well as two other small fountains connected to a brickwork tank in the principal square of Roccarainola. Nowadays, only the little fountain on the hill, near the qanat entrance, continues to slowly supply water. Since intensive urban development surrounded the qanat, leading to heavy groundwater pollution, the water of the small fountain is not drinkable. D'Avanzo (1943) stated that the Roccarainola qanat was probably not constructed by the Romans but during the Middle Ages, in order to supply the nearby Norman castle. D'Avanzo (1943) based his statement on the fact that in Roman times, Ages there were no important cities near the area of the qanat. Moreover, the nearby cities of Avella and Nola were supplied by other aqueducts (De Feo and others, 2009).

The Roccarainola qanat is composed of two principal branches. One branch is developed in a north-northeast direction ("North Branch"; Figure 5.10), while another branch goes east ("East Branch"). A secondary branch starts from the "East branch" and heads in a northeast direction

Figure 5.10 Typical cross section of the North Branch of the Roccarainola qanat.

Source: G. De Feo and others, 2009.

("Springs Branch"). The underground tunnel has a total length of 786 meters, with a depth of 9 meters (De Feo and others, 2009).

5.10 Afghanistan

In Afghanistan there are still some active qanats (karezes) which are used to irrigate farmlands. The annual groundwater discharge throughout the country from pumped wells and arhads (probably shallow dug wells) is estimated to be around 340 million m^3/year. The annual groundwater flow from natural springs and qanats is respectively estimated to be around 940 million m^3/year and 1,770 million m^3/year. The total number of qanats is 9,370, of which e active qanats amount to 5,984. More than 36% of the qanats have dried up and the discharge of the remaining ones has reduced significantly. The average length of a qanat in Afghanistan is 1.74 km and the total length of main branches of karezes in the country is estimated to be 16,374 km. The average discharge of a qanat is 6 liters per second and the total cumulative discharge of qanats in the country is estimated to be 55,800 liters per second.

Karezes/qanats are found in 19 provinces of Afghanistan: Kapisa, Khost, Kunar, Laghman, Logar, Nangarhar, Paktika, Paktia, Parwan, Urugan, Wardak and Zabul. These provinces are located in 3 main river basins of Afghanistan, one of which is the Harirud-Morghab, Hilmand and Kabul River Basin. No reports on the existence of karezes in the remaining two river basins of Northern and Amu Darya basins are available in the recent survey in 2003 but there are reports (statistics published in 1980) on the presence of karezes in the provinces of Samangan and Jawzjan and Balkh, and also in Nimroz, Ghor, and Faryab (Table 5.1, Figure 5.11).

The Ghaznaweed civilization, around 1000 years ago, played an important role in the expansion of irrigation systems with surface and groundwater sources in the region. The Ghaznaweed built many dams along the Hilmand and Ghazni Rivers and constructed thousands of karezes in the southern, western and eastern parts of their territory. Invasion by Alaudin Jehansooz led to the destruction of most of the dams and karezes in the region. During later eras, of the Ghories, Timories and Abdalies, rebuilding of destroyed karezes and building new karezes started, and their use became prevalent. They played a key role in the maintaining, operation and construction of new karezes.

Table 5.1 Number of qanats in each province in 2003

1	BADGHIS	31
2	FARAH	164
3	GHAZNI	2,112
4	HILMAND	146
5	HIRAT	390
6	KABUL	932
7	KANDAHAR	746
8	KAPISA	130
9	KHOST	88
10	KUNAR	6
11	LAGHMAN	21
12	LOGAR	279
13	NANGARHAR	300
14	PAKTIKA	1,476
15	PAKTYA	668
16	PARWAN	29
17	URUZGAN	217
18	WARDAK	740
19	ZABUL	895
	Total	9,370

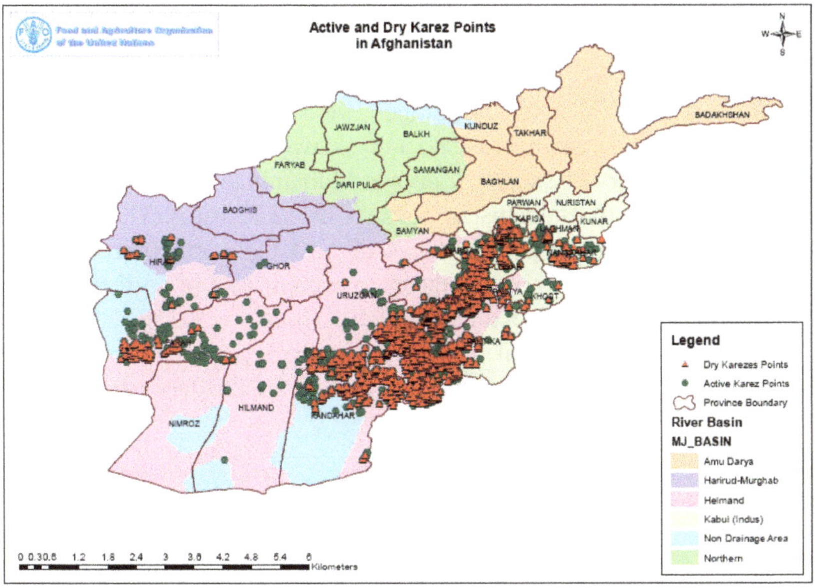

Figure 5.11 Distribution of qanats throughout Afghanistan.

Source: Sayed Sharif Shobair, 2008.

Over the last few decades, years of war and turmoil in Afghanistan took a heavy toll on many karezes. War has direct and indirect effects on karezes: the Mujahedeen/fighters used to take shelter in the karezes or used them for ambushes; therefore karezes became a target of Soviet and other invaders. Also, during war, many villagers took refuge to the neighboring countryside and nobody was available for the routine maintenance of karezes for many years, this has caused the destruction of the tunnels and wells. The recent persistent droughts, introduction of simple and easy drilling machinery and ineffectiveness of law and regulation enforcing agencies has caused drying up of hundreds of karezes. Due to continuous drought and low rainfall, the groundwater recharge has reduced significantly and accordingly the flow from the karezes has diminished, and many have dried up completely. This dramatic decrease in the discharge of karezes forced people to drill deep wells in the vicinity of the karezes, causing further drying up of karezes and also creating disputes among the local communities (Sayed Sharif Shobair, 2008).

5.11 Pakistan

The only province of Pakistan which benefits from qanats is Balochistan, but over the last decades, a lot of pumped wells have been drilled there (Figure 5.12). Kahlown considers Pakistan as two areas – northern and southern, in his inventory of qanats (Kahlown Muhammad Akram, 2008).

The northern zone of Balochistan consists of 12 districts. The Provincial Irrigation and Power Department has divided the northern zone into 7 irrigation divisions. The number of qanats in each district of northern zone ranges from 4 (Quetta) to 168 (Killa Saifullah). There are 563 qanats in the entire northern zone, with a total flow of 8,064 liters per second. There is wide range of flow from these qanats, from 53 to 5,222 liters per second.

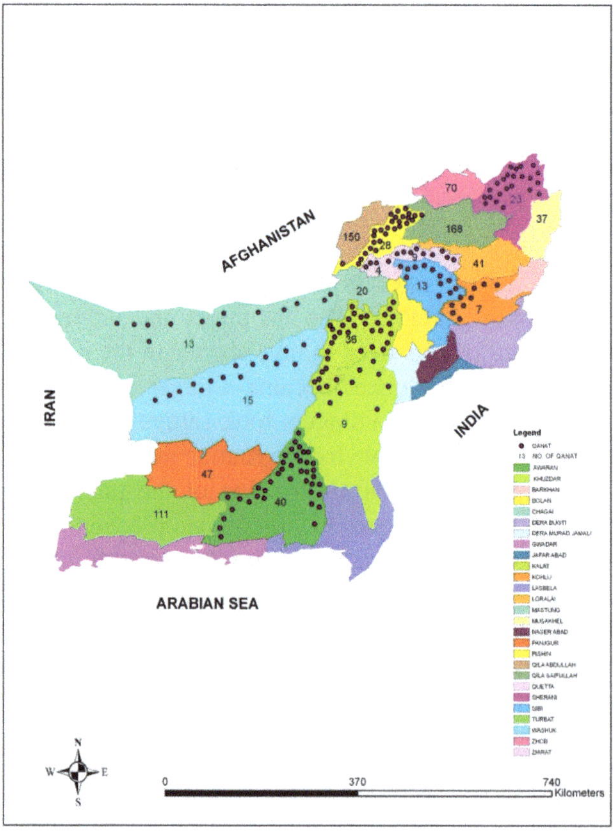

Figure 5.12 Location map of qanats in Balochistan, Pakistan.

Source: Kahlown Muhammad Akram, 2008.

Table 5.2 Inventory of qanats in Northern Balochistan

S. No.	District	No. of Qanats	Discharge Liter Per Second (lps)	Area Under Cultivation (hectare)	Beneficiaries (Nos.)
1	Quetta	04	53	26	60
2	Chagai	13	364	71	168
3	Pishin	28	616	1038	3677
4	Killa Abdullah	150	5222	5560	35,313
5	Sibi	13	1106	86	546
6	Ziarat	09	180	85	355
7	Loralai	41	853	930	636
8	Musa Khel	37	641	860	509
9	Zhob	70	1813	2255	9128
10	Sherani	23	457	571	2350
11	Killa Saifullah	168	2706	2984	3635
12	Kohlu	07	238	271	371
	Total	563	14,249	14,737	56,748

The area commanded by each qanat ranges from 240 to 3738 hectares, and the total area under cultivation by the qanats is 12,465 hectares. There are 13,474 farmers benefiting from the qanats in the northern zone (Table 5.2).

The southern zone in general is inhabited by the Balochs and consists of 7 districts. The provincial Irrigation and Power Department has divided the southern zone into 4 irrigation divisions. In the Southern zone, 278 qanats with the total discharge of 8,062 liters per second are operating. The number of qanats in the different districts ranges from 9 to 111. Their discharge ranges from 200 to 3,889 liters per second, and the area under cultivation from each qanat ranges from 240 to 3,788 hectares. The beneficiaries on each qanat range from 258 to 7,117[6] (Table 5.3).

5.12 Sultanate of Oman

In Oman, the qanat (locally called aflaj/falaj) is rooted in history. Gred Weisgerber who has studied the qanats of Oman from the archeological point of view writes that "Aflaj are not static installations, as they were described so far. They have to follow the natural consequences of their existence. They are dynamic constructions which always need intensive labor adapted to the

[6]Qanat in Afghanistan, Pakistan & Iran, Edited by Semsar Yazdi and Labbaf Khaneiki, Report of Pakistan by Dr. Mohammad Akram Kahlown, Report of Afghanistan by Sayed Sharif Shobair, Report of Iran by A. A. Semsar Yazdi, M. Labbaf Khaniki, S. Askarzadeh.

Table 5.3 Inventory of qanats in Southern Balochistan

S. No.	District	No. of Karezes	Discharge (lps)	Area Under Cultivation (hectare)	Beneficiaries (Nos.)
1	Khuzdar	09	269	1126	258
2	Awaran	40	960	3738	975
3	Kech (Turbat)	111	3889	3528	7117
4	Panjgur	47	1679	1512	3066
5	Mastung	20	200	754	689
6	Kharan	15	200	240	502
7	Kalat	36	865	1568	867
	Total	278	8062	12,466	13,474

changing circumstances. In doing surveys, one should follow the course of the qanats to determine the areas of prehistoric settlements connected with the qanats. At the moment we are not able to say when this system of water tapping and transport water was introduced into Oman. The system was introduced in Oman at least by the first half of the first millennium BC, an age earlier than the Hellenistic report for Iran, and possibly earlier than the Achaemenian introduction once supposed ... the astonishing example of al-Maysar with its downstream moving settlements as an impact of changes in the balance of groundwater in response to water exploitation in Oman and elsewhere remained an unreported human reaction to changing conditions of life ... it is more than probable that those people who after the Bronze Age started a new flourishing second era of settling and cultivating Oman in the Iron Age by introducing the water procurement by the qanat system, had to build those fortresses all over Oman. The connection between Iron Age settlement and fortresses and the qanat system is obvious" (Weisgerber Gred, 2003).

Apart from the historical value of the qanats of Oman, from the economic point of view this system is still of great importance. Qanats are still playing a vital role in agricultural systems in Oman. Due to the concern of the beneficiaries over qanats and the Omani government's support, fortunately many qanats in Oman are still flowing.

According to a report published by the Ministry of Regional Municipalities, Environment and Water Resources of Oman, this country has 4,112 qanats of which 3,108 are in active use and the rest are beyond utilization (Aflaj Inventory Project Summary Report, 2001).

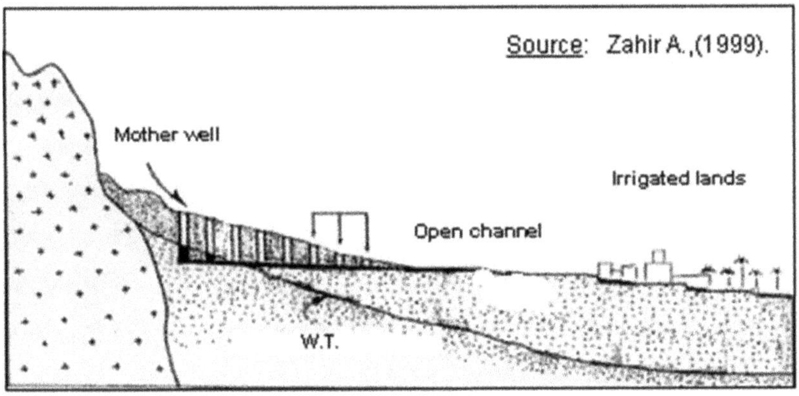

Figure 5.13 Daudi falaj.

In Oman, the qanats are called "aflaj" which are classified into three groups as "Daudi", "Aini" and "Ghaili", depending on whether they utilize shallow or deep groundwater or surface water.

Daudi: A qanat that draws water from significant subterranean depth is termed "daudi" (Figure 5.13). Usually, the flow continues without being significantly affected by the changes of season at the locale. A falaj is termed daudi when at least one subsidiary branch conforms to the definition.

Aini: A qanat that draws water from one or more natural springs is termed "aini" (Figure 5.14). The water flows from springs into the falaj channel, which transports it to agricultural lands. A falaj is termed aini when at least one subsidiary branch conforms to the definition and when there are no branches that conform to the definition of a daudi falaj. For example Falaj Al-Jeela is considered an aini falaj located in Al-Jeela village, Willayat Sur. It is fed by wadi Shab in a rough mountainous area. It is the main water source for the village with a total length of its open channel at 161 m, which starts from the wellhead and ends at the water catchment basin. The channel is adjacent the mountain at a lofty height. It originates from a solid rocky limestone area. It has pure water where the electric conductivity is 378 microsiemens/cm only, with a pH value of 7.87 and water temperature of 29° Celsius. It flows all year round with negligible impact As a result,of increase and decrease of water level, where its average flow reaches 0.044 m^3/s[7].

[7]http://www.mrmwr.gov.om/index.asp?id=110

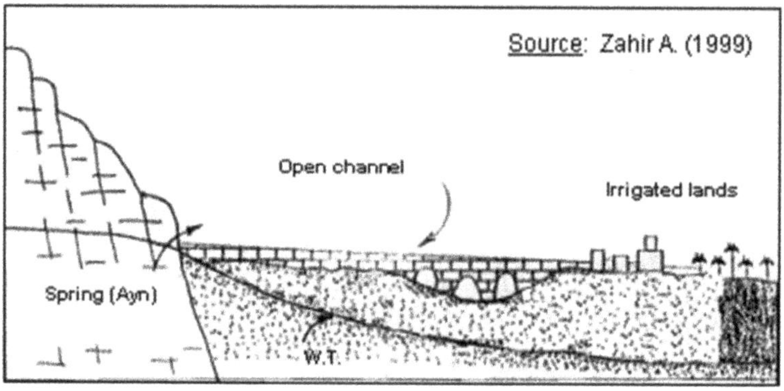

Figure 5.14 Aini falaj.

Ghaili: A qanat that draws water from a wadi, from water flowing on the surface of the earth, or from subsurface water close to the surface of the earth in wadi channels or at the foot of mountains is called "ghaili". A falaj is termed "ghaili" when it depends only on a subsidiary branch or branches that conform to the definition (Figure 5.15).

Approximately 23 percent of Oman's qanats are daudi, and aini and ghaili types make up respectively 28 and 49 percent of the countries' qanats. Most of the dead qanats in Oman are of the ghaili type (54%), while dead daudi and aini qanats are 31% and 15% respectively of the total. The area dependent on qanat water in Oman is 26,498 hectares, 66.3% of which is cropped, 3.7% is

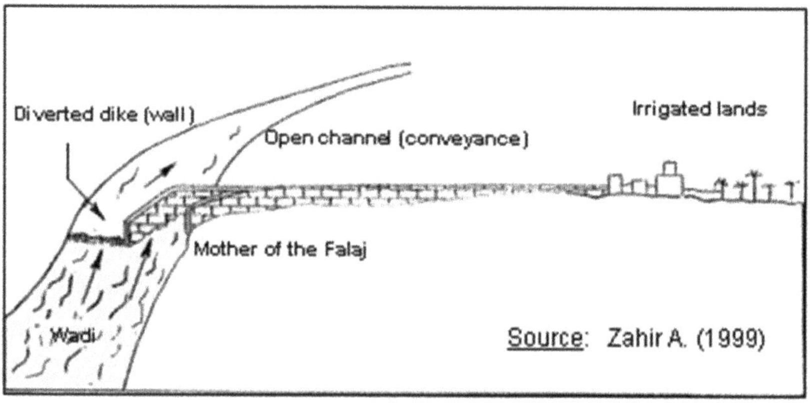

Figure 5.15 Ghaili falaj[8].

[8]http://www.unep.or.jp

developed uncropped, and 30% is undeveloped. The total annual demand for qanat water in Oman is 460 million cubic meters. The electrical conductivity of Oman's qanats ranges from 115 $\mu s/cm^2$ to 16,700 μ/cm^2. Chemical tests by the water quality laboratory belonging to Ministry of Municipalities, Environment and Water Resources show that the water of 2,762 qanats is suitable for cultivation of all crops.

The Ministry of Environment and Water Resources of Oman has a section named Directorate of Water Resources Management that favors qanats and their rehabilitation. In sum, qanats and their roles in sustainable development are the center of Omani government's attention, and are supported by the government. The government of the Oman has formulated appropriate strategies to manage water demand and its use with the objective of achieving a sustainable water balance. The government administers a sound well permit system for drilling or deepening of wells with penalties in case of violation. Well field protection zones have also been established for key water supply schemes to prevent pollution and over-pumping. Public awareness campaigns are an important component of the programme and accompany all new legislation as well as raising general water conservation issues through public media (Al Marjeby Aley, 2002).

Another issue which is of great importance in terms of water consumption management in Oman is the indigenous methods for the division of falaj water. From early on, Omani people have practiced an elaborate water management and water division system. Al-Ghafri writes: "the most common method of distributing water in aflaj of Oman is to divide the irrigation rotation (dawran) into several days; each full day is divided into 48 athars. Each athar is then divided into smaller subdivisions. Traditionally, farmers have used sundials in the daytime, and stars at night for justifying water shares on the field. Due to the change of the day and night lengths around the year, the practical length of athar also varies. To overcome this variation, farmers have adapted many ways to insure equitable distribution of water among them... it can be concluded that the development of aflaj should not be considered only hardware, structures and materials but also software, management and the water-share system" (Al-Ghafri, 2002).

As mentioned in Chapter 3; Social Aspects, Harriet Nash has also studied the water allocation based on time calculation by stars in Oman, which sheds a light on the cultural value of aflaj systems in that country (Nash Harriet, 2007).

Figure 5.16 Falaj Al-Khatmeen in Oman.

Source: http://whc.unesco.org

In closing, recently five qanats of Oman have been registered in the UNESCO World Heritage List, these qanats are Falaj Daris and Falaj Al-Khatmeen in Willayat Nizwa (Figure 5.16), Falaj Al-Malaki in Willayat Izki, Falaj Al-Mayssar in Willayat Al-Rustaq and Falaj Al-Jeela in Willayat Sur[9]. These five qanats represent the other qanats in Oman, which are the symbol of sustainable development (Table 5.4).

One of the most recent researches on Aflaj has been conducted by Feirouz Megdiche as her PhD dissertation. This research focuses on the role of Aflaj in configuring the landscapes of hydraulic societies in Oman (Megdiche, 2018).

5.13 Republic of Azerbaijan

There are an estimated 1,500 qanats in Azerbaijan of which 885 are still operational (Table 5.5). Although their total discharge is not that considerable, some communities still live off the water coming out of these qanats. One of the regions of Azerbaijan is Nakhchivan Autonomous Republic that houses 407 qanats with a total discharge of 1,330 liters per second, irrigating some 3,000 hectares of cultivated land. Since 1960, some of the qanats have dried up, mainly because of negligence, and As a result,the number of active qanats in Nakhchivan dropped to 288 (Musayev Arzu, IOM).

[9]According to the website of UNESCOUNESCO world heritage at the URL http://whc.unesco.org/en/list/1207/ Also according to the website of Ministry of Regional Municipalities and Water Resources, Sultanate of Oman at the URL http://www.mrmwr.gov.om/index.asp?id=74

Table 5.4 Maximum and minimum discharge for some selected aflaj in Oman

	Zone	Falaj Name	Falaj Type	Maximum Discharge (L/s)	Minimum Discharge (L/s)	Micomhos/ Sec	Temperature (°C)
1	Interior	Daris	Daudi	2,160	40	480	31.6
2	Interior	Alghantaq	Daudi	350	10	370	33.6
3	Interior	Almythaai	Daudi	620	3	756	28.2
4	Interior	Almalaky	Daudi	780	10	659	35
5	Interior	Alsamady	Ghayli	150	60	806	28
6	Interior	Alkali	Ghayli	100	34	1,003	24
7	Interior	Blfaai	Ghayli	80	30	850	30.8
8	Interior	Alhamam	Daudi	20	3	799	39
9	Eastern	Alkamil	Daudi	120	30	690	33
10	Eastern	Alwafi	Daudi	150	40	616	33
11	Eastern	Alfarsakhy	Daudi	300	30	1,122	32
12	Eastern	Daraiz	Daudi	250	20	1,056	33
13	Eastern	Alghaby	Daudi	220	20	990	34
14	Eastern	Albahmady	Daudi	270	30	800	33
15	Al-Dhaira	Alsald	Ghayli	450	30	751	29.5
16	Al-Dhaira	Salalah	Ghayli	230	4	918	26.7
17	Al-Dhaira	Daraiz	Daudi	600	20	500	32.8
18	Al-Dhaira	Almabooth	Daudi	220	10	1,196	32
19	Al-Dhaira	Aloolaw	Daudi	210	20	952	40.5
20	Al-Batina	Ain Alkasfah	Ainy	170	40	1,068	44.9
21	Al-Batina	Almaysar	Daudi	1,390	30	624	32
22	Al-Batina	Alkabail	Daudi	140	10	941	32.8
23	Al-Batina	Aloohi	Daudi	250	10	575	31.6

Source: Al-Suleimani Z., 1999.

Given the technical and historical value of the qanats in Azerbaijan, measures were taken to mitigate the pitiable situation of this hydraulic legacy (Figure 5.17). One of the organizations that took the lead in rehabilitation of qanats in Azerbaijan was the International Organization for Migration (IOM).

In 1999, upon the request of the communities in Nakhchivan, taking into consideration the needs and priorities of the communities, especially women as the main beneficiaries, IOM implemented a pilot program to rehabilitate the qanats. In 2004, the Government of the Nakhchivan Autonomous Republic also intervened by establishing a qanat department, which aims to institutionalize the functioning, management and protection of qanat systems. With the support of IOM, the department has renovated several qanats in Nakhchivan. The Kahriz Resource Center, affiliated to the qanat department was established in 2006 and is promoting this approach in Nakhchivan as well as in Azerbaijan mainland. The center's activities cover

Table 5.5 Statistics of qanats in different regions of Azerbaijan

	Regions	Number	Water Length, Discharge, L/sec	Length, km	Number of Wells
1	Gazakh	8	97.0	17,710	530
2	Tovuz	5	297.0	12,719	361
3	Shamkir	29	842.0	14,926	734
4	Ganja	103	2500.0	166,829	7062
5	Goranboy	20	261.0	21,644	693
6	Barda	45	1428.0	40,051	2024
7	Yevlakh	4	150.0	3,759	222
8	Tartar	2	54.0	1,143	49
9	Agdam	105	2040.0	112,424	3908
10	Agjabedi	68	1618.0	124,278	4630
11	Fizuli	71	603.75	37,830	1491
12	Jabrail	111	1099.5	59,311	2234
13	Montain Garabkh	52	134	20,181	887
14	Sharur	99	203.6	21,974	903
15	Babek	86	1365.5	31,714	1058
16	Shabuz	4	100.0	1,042	29
17	Abragunus	5	50.0	1,533	43
18	Ordubad	68	528.5	31,940	992
	Total	885	135,35 9	721,008	27,850

Source: Archive of Azer State Water Project Institute.

the areas of documentation, data collection, gathering and sharing of technical information.

Taking into account the important role the kahriz system is playing in the region, a research project was carried out in the wake of an agreement between IOM and ICQHS International Center on Qanats and Historic Hydraulic Structures (Figure 5.18). The focal point of this agreement was an assessment of the situation of kahrizes as well as the efficiency of the procedure conducted by IOM herein. IOM has well-recognized the important role that kahrizes can play in rural communities. Given that most of the rural areas are in the grip of water scarcity, the villagers are prone to migrating to the larger towns and cities, bringing about many social and economic side effects. Drilling pumped wells to meet the demand of the farmlands is not an appropriate remedy, because the experience of some similar countries shows that resorting to pumped wells may lead to the depletion of groundwater, earth subsidence, groundwater salinity and eventually the annihilation of groundwater resources. This research led to the construction of new qanats in Nakhchivan. IOM built on the research conducted by ICQHS

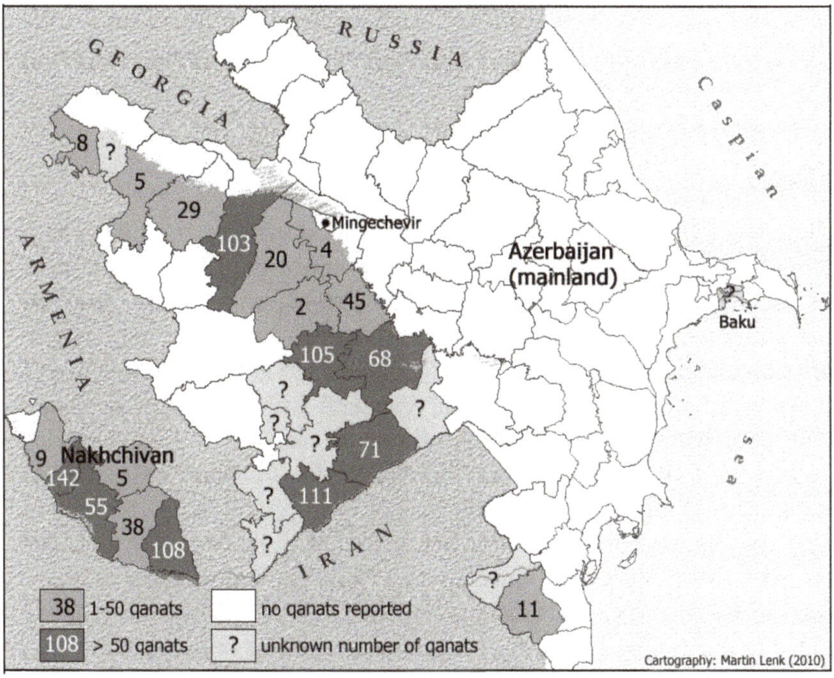

Figure 5.17 Distribution of qanats in Azerbaijan.

Figure 5.18 IOM experts examining a kahriz in Nakhchivan.

and managed to construct new qanats in Nakhchivan in the 21st century. This initiative could bridge the egregious gap between past and future at least in the field of groundwater exploitation.

5.14 Iraq

According to Lightfoot "The dating of qanat (called karez in Iraq) anywhere is notoriously difficult, even where material evidence is plentiful, and must rely on circumstantial evidence around a karez or historical details from an adjacent settlement clearly linked to a karez through its exit tunnel, cistern, or distribution canal. Even these lines of evidence are scarce or non-existent for the karez of northern Iraq and only general inferences can be made regarding their historical introduction, diffusion, and use... Oral histories often report local karez to be 100–350 years old. In many cases this is contemporary with the age of the village. For example, 22 karez lie around the village of Goradem in the Sharbazher district of Sulaymaniyah, a village that has existed for 280 years (since about 1730). The oldest karez here are presumed to be this age, but others were built in the 1920s–1930s (and 2 or 3 added in the 1960s–1970s) by Iranian qanat masters brought to the area from Mariwan, Bane, and Bukan. Many karez in Iraq are simply reported to be "very old," meaning before living memory or the stories of fathers and grandfathers. The modern Iraqi historian Abdul Razzak Al-Hassani reports that the karez in Kurdistan are "very old" and in very ancient times they allowed Erbil to continue developing and thriving, because invaders could not control Erbil or destroy the karez as invaders had done to water sources in the past in Babylon, Ur and other southern cities of Iraq. It is clear that most karez in Iraq were constructed "long ago" but not much more can be said with any accuracy when speaking about a specific karez. A few can be dated more precisely using old documents that mention the karez. Several owners of karez reported that they have documents that attest ownership, and some of these documents (mostly 19th–20th century papers) record new construction" (Lightfoot Dale, 2009). Lightfoot attributes the ancient qanats in Iraqi Kurdistan to the time of Christians, saying "many karez are said to come from the "Gaurkare" period; a reference to the Christian era, meaning pre-Islamic. "Gaur" means "Christian" here (people in Iraqi Kurdistan call Jews, "Jews," but call Christians, "Gaur"). "Kare" means "work," so Gaurkare refers to Christian works, or simply to the Christian and pre-Islamic era of AD 1 to 632. The Parthian and Sassanid Persians were not Christians, but their works were added to the landscape during this period and may be referred to as Gaurkare. Regardless of religious affiliation, it is clear that Kurdish people, before Islam, knew how to dig karez and constructed them for different purposes" (Lightfoot Dale, 2009). It is worth noting that the word Gaur does not mean Christian, in contrast to Lightfoot's account. This world has repeatedly been used in the 1,300 years history of Persian literature: Gaur or Gabr has nothing to do with Christians but means Zoroastrian, so Lightfoot's speculation no longer contradicts the historical reality.

At present, most of the qanats in Iraq are concentrated in the region of Kurdistan, and due to the years of conflict in the region there is no accurate data on qanats available. So we have no exact idea even on the numbers of qanats running in the region of Kurdistan.

Recently, Dale Lightfoot has documented 683 infiltration karez[10] throughout the northern governorates (Dohuk, Ninewah, Erbil, Kirkuk, and Sulaymaniyah). The karez in this region have been adversely impacted by drought and excessive well pumping. Almost 40% of karez documented – and 70% of those that were still flowing five years ago – have been abandoned since the onset of drought in 2005. As a result,of this decline, over 100,000 people have been forced to evacuate their homes since 2005. His study identified 116 karez that were still being used in summer 2009, but all have diminished flow, placing an estimated 36,000 people at risk of displacement (Lightfoot Dale, 2009). Many qanats in Kurdistan suffer the most from negligence: the build-up of sediment in their galleries indicates that these qanats have long been abandoned. Sometimes the sediment is up to 1 meter thick, and the gallery has not been cleaned out for tens of years. UNESCO's Iraq Office has recently allocated a fund for the rehabilitation of qanats in rural areas where the qanat can still play an important role. ICQHS cooperated with UNESCO to study some sample qanats to better figure out how to rehabilitate and maintain them. Based on the study done by ICQHS, UNESCO's Iraq Office singled out the karez of Shekh Mamudian in Northern Iraq as a pilot project (Figures 5.19 and 5.20). Eventually UNESCO completed the refurbishment of this karez, returning a source of water to the small community of Shekh Mamudian. The karez had stopped flowing in 2007. A feasibility study done by ICQHS in July 2009 concluded that drought had caused the water table to recede only to a depth of 1.5 meters below the karez tunnel. Work focused on deepening the tunnel and extending

[10]Infiltration karez means a karez that cuts through a saturated zone underground and drains out the seepage of groundwater. This kind of karez is different from those which get water from a natural spring and only conveys water to somewhere else on the surface.

Figure 5.19 Pumping water from the borehole to gauge the potential water discharge of the kahriz Shekh Mamudian during the study done by ICQHS.

Figure 5.20 ICQHS expert at the bottom of the third shaft well to examine the borehole and take a water sample.

the karez outlet by 300 meters, using traditional methods. A ribbon-cutting ceremony took place at the village on Friday, 29 January 2010, to inaugurate the works[11].

[11]for further information goa on the UNESCO website at http://www.unesco.org/en/iraq-office/single-view/news/unesco_completes_kahrez_restoration_in_northern_iraq/back/9623/cHash/e7a38132ee/

5.15 India

We have no idea about the exact number of qanats all over India. All we can say is that there are some 5000 qanats in the states of Kerala and Karnataka. However, those qanats are referred to as "suranga" which denotes a special qanat system which is relatively small, owned by the petty farmers and used at a very local level. Apart from suranga, we could track down a different type of qanat in the state of Karnataka, following one of the authors' mission to India. This type of Indian qanat is called karez; the same name used in eastern Iran, Afghanistan and Pakistan. One of the good examples of the Indian karez is located in the city of Bidar, north of Karnataka.

In historical city of Bidar in India, water has always been a subject on which the engineers could have exhibited their genius for such hydraulic structures as embankments, water towers, cisterns, canals, karez (qanat), etc, and a possibility for the kings to show off their majesty through building splendid fountains, pools, waterfalls, and brooks. Water used to be the cornerstone of a civilization which had to deploy a great deal of ingenuity in adapting itself to the fickle climatic conditions of this region where monsoon pours down torrential rains and dry seasons dehydrate the land. In the course of history, these people had to adjust to water fluctuation on one hand, and to live up to each other's demand for water on the other hand.

In such a context, karez (qanat) system in Bidar is unique and has resulted from a systematic relationship between human and its environment. Karez and water reservoir combine as a unique structure. The wells have been dug in a relatively large size as well as below the level of karez tunnel in order to store water when karez water dwindles or dries up; a hybrid of karez and water reservoir. The amazing characteristics of qanat system in Bidar exemplify the human's historical relationship with water resources.

North of Karnataka including Bidar and Bijapur is overwhelmed by a rich water history, which is anchored in its environmental conditions (Figure 5.21). In Bidar, groundwater can be stored in a layer of laterite which varies in thickness from 10 to 30 meters. A layer of impermeable basalt underlies the layer of laterite, which creates a situation suitable for a shallow aquifer. The porous laterite can easily soak up the runoffs during the rainy seasons and also lose its water again easily during dry seasons. Hence,the depth to groundwater level varies from 2.81 to 17.90 meters below ground surface in May, and this number abruptly changes to between 0.40 and 13.09 in November. This sharp fluctuation in the water table drove the inhabitants of this region to come up with many innovative techniques to better tackle

Figure 5.21 The well number 3 of the qanat system in Bidar.

water problem. All these techniques revolve around the fact that water influx should be harnessed and stored during rainy seasons to be used during dry seasons when natural precipitation cannot meet the water demand. It is not exaggeration if we call Bidar a big water museum which reflects the best example of human capability in adapting themselves to the environmental conditions.

Qanat or karez system in Bidar is unique and has resulted from a systematic relationship between human and its environment. It cannot be completely ruled out that the early people of Bidar adopted the idea of qanat technology from Persians especially at the time of Bahmani dynasty who were on very friendly terms with Iranians. Even if we accept it, this idea has been tailored to the geographical and climatic conditions of Bidar by incorporating many local elements into the so called qanat technology. Therefore this type of karez is specific to this region and is an intrinsic element of this cultural landscape. In a nutshell the main characteristics of karez in Bidar can be listed as follows:

1. Karez and cistern combine as a unique structure. The wells have been dug in a relatively large size as well as below the level of karez tunnel in order to store water when karez water dwindles or dries up. In fact this type of qanat exhibits a hybrid of karez and water reservoir.
2. Tools and techniques used in karez construction were specific to Bidar whose conditions did not allow the workers to apply Persian tools for example. A Persian windlass could not be installed for a square well with such large dimensions of 5 by 5 meters. Geological formations of the region also gave rise to a set of special tools and equipment.
3. Architectural value of Bidar karez system is unique. Laterite gave the early builders an opportunity to consider the aesthetic aspect of karez and deploy the architectural heritage of the region even in karez. Almost in every version of qanat, what the workers usually do is to devise an engineered tunnel to tap the aquifer. But in Bidar, they were not only concerned about the engineering aspect of karez construction, but also envisioned architectural possibilities by building beautiful stone walls, elaborate arches cut out of laterite, etc.
4. Gradient of Karez tunnel does not follow the surface slope, which results in direction of surface streams opposing that of karez flow.
5. The karez wells have been all dug square in shape. This shape helped them to find the right direction underground where the early workers wanted to dig a tunnel from a well to another. Also this shape was suitable for mining and cutting out cubic stone blocks for other constructions on the surface (Labbaf Khaneiki, 2017).

5.16 A Review of Conferences on Qanats Held in the World

All the conferences held on qanats so far indicate that this ancient hydraulic system has been a subject of close attention by many scientists and authorities in recent years, so giving hope for future. This part of the book examines all conferences and meetings on qanats which have been held from 1980 to 2009. The seminars which have been held in Mashhad (Iran), London (UK), China, Yazd (Iran), Paris (France), Spain, Muscat (Oman), Luxemburg, Gonabad (Iran), Kerman (Iran) and Nakhchivan (Azerbaijan) analyze this traditional know how from the geological, anthropological, archeological, civil engineering, sociological, hydrogeological, irrigational, legal ..., etc. points of view. Assessment of all these seminars and their declarations as well as the outcomes not only enable us to find the main results of the research, but enable us to plan future scientific activities.

The technology of qanat is an ancient water extraction system in perfect harmony with nature. without disturbing the aquifer. As mentioned, attention to indigenous know-how as an environment friendly solution is catching on in the world, bringing about numerous conferences on qanats since 1980. In this part, we try to survey these occasions closely and specifies the available social turbulence concerning this hydraulic system according to the articles presented and the declarations issued in order to decide on the qanat safeguarding priorities. Dissemination of such essentials can pave the path so that the next steps can be taken for the development of sensible strategies in order to preserve these ancient hydraulic systems. The concerned researchers may benefit from the viewpoints raised by archeologists, geologists, civil engineers, geographers, experts of social sciences and economics, hydrogeologists, legal experts, etc. Here we take up the conferences chronologically.

5.16.1 Seminar on the Renovation and Rehabilitation of Qanats, Mashhad 1981

This conference was held in Mashhad (Iran) from 1st to 7th of July 1981, sponsored by the Budget and Planning Organization of Khorasan to seek remedies for qanat rehabilitation countrywide. Experts from most of the provinces provided information on their activities and achievements concerning the rehabilitation of qanats to the secretariat. The lectures about the Iranian qanats were presented under the following titles:

1. A report raised by the Budget and Planning Organization of Khorasan about the qanats of that region
   - (i) Historical records and importance of qanats in Khorasan
   - (ii) Qanat discharge importance in Khorasan
   - (iii) Operation, preservation and reclamation of qanats in Khorasan
   - (iv) Investment and the achievements in operation, preservation and reclamation of qanats in recent years
   - (v) Programs for the preservation and reclamation of qanats in Khorasan
2. Recognition of qanats and their role in sustainable development of groundwater resources, by Arash. Kuchakpur.
3. Artificial recharge of the aquifer by Mahdi Parsi
4. Geographic location and expansion regarding traditional legal bounds of qanats and the present evolution of this phenomenon by Hassan Ali Zadeh

5. Floods in the plateau of Iran and damage to qanats, Parviz Nilufari
6. A report by the Budget and Planning Organization of the province of East Azarbayejan
7. A report by the Budget and Planning Org. of the province of Fars.
8. Educational revolution and the establishment of the qanat training center, Hassan Abrishami
9. A report by the Budget and Planning Org. of the province of Yazd
10. Qanat and vegetation, Roushan
11. Qanat as the source of production in rural areas, Mohammad Taghi Farvar
12. A report by the Budget and Planning Org. of the province of Esfahan
13. A report by the Budget and Planning Org. of the province of West Azarbayejan
14. Qanat Ecosystem, by Morteza Honari
15. A report by the Budget and Planning Org. of the province of Sistan and Baluchestan
16. A report by the Budget and Planning Org. of the province of Kerman
17. Reclamation and amelioration of the Qanat discharge by Hassan Iran Manesh
18. Qanats' role in water supply for desert areas by Hormoz Pazush
19. Alleviation of Qanats' insufficiencies with the rural residents' collaborations Mohammad Ashtiani
20. A report by the Budget and Planning Org. of the province of Semnan
21. A report by the Budget and Planning Org. of the province of Lorestan

The outcome of this seminar was a declaration in seven articles as follows:

1. All the participants agreed on the necessity of preservation of qanats and their undeniable role in regeneration of rural areas and production increase.
2. Abandonment of numerous qanats in Iran and decrease of discharge caused the disintegration of the traditional producing systems of rural cooperatives, causing the migration of the dwellers to metropolises, which has disturbed the rural economy. Consequently, qanats play a vital role in Iran's domestic economic prosperity.
3. Traditional system of groundwater use alleviates Iran's technical dependency.
4. Five working groups dealing with technical policy, economic, social, educational and irrigation affairs are to be formed, to be in charge of the problems raised and their solutions.

5. Since inspections and pursuing qanat concerned issues last long, provincial qanat committees continue the activities performed by the working groups, harmonizing well with the qanat rehabilitation headquarters.
6. Preservation of vegetative cover where there are qanats should be considered precisely because proper vegetation can prevent hazards of runoff and increase the recharge of water into the aquifer.
7. All the researchers and scientists are invited to carry on their qanat investigations and disseminate the outcomes for looking into qanat concerned issues more closely.

5.16.2 Qanat, Kariz and Khatara, Traditional Water Systems in the Middle East and North Africa, London 1987

This symposium was held in London in 1987 reviewing qanats from historical and geographical aspects. The proceedings of this conference were published by MENAS publications (MENAS Press). Regretfully, we do not have full information about the organizers, participants and the main objectives followed by the conference.

5.16.3 International Conference on Karez Irrigation, China 1993

This conference took place in Urumqi, China in 1993 and was jointly sponsored by the Chinese Committee of Man and Biosphere (MAB) and the Xinjiang Institute of Biology, Pedology and Desert Research Academy of Sciences, under the auspices of the Academy of Sciences and Xinjiang Uygur Autonomous Regions. A joint research project performed by the Japanese University of Meiji, Xinjiang Institute of Biology, Pedology and Desert Research Academy of Sciences of Xinjiang from 1988 to 1990 caused great strides to be made for the observance of this occasion. The whole research was performed on qanats in Xinjiang (China) bringing about very fruitful outcomes. More than 80 researchers and scientists coming from ten nationalities brainstormed on different features of this ancient know-how regarding the origin, distribution, future and management of qanats. The presented articles in this conference were the followings:

- "Study on Xinjiang Karez" by Wang Hoting.
- "Comparative studies on the formation of qanat water system" by Kobori Iwao.
- "Re-expound the source and development of Xinjiang Karez" by Huang Shengzhang.

- "The relation of the Karez systems of Xinjiang and the central plains of China" by Ciapan.
- "Studies on Xinjiang Karez" by Xinjiang Water Conservancy Bureau and Xinjiang Hydraulic Engineering Society.
- "The preliminary studies on the origins of Karez in Turpan Basin, Xinjiang" by Fanzili.
- "The study of historical geography on the origin of the Karez of Turpan" by Wang Shouchun.
- "The source and utilization of Xinjiang Karez" by Liu Chifei.
- "The conditions of the formation of Karez in Xinjiang and the theories about their origin" by Zhang Xiru.
- "The present situation of the Karez in Turpan Basin" by Huang Zhixin.
- "Rock drawing – the testimony for the research on the Karez origin" by Alim Niyaz.
- "The origin of the Karez in Turpan – reason for its spread" by Sunao Hori.
- "Discussions on the research textual origin of Karez in Xinjiang, China" by Liang Yide.
- "The origin and development of Karez in Turpan Basin" by Ha Yun-chang.
- "The beginning and end of Karez digging in Pishan district of Xingiang" by Yu Wenru.
- "The ancient irrigation works as seen in the Xinjiang archaeological investigation" by Wang Binghua.
- "General view of the Karez culture as seen from the excavated documents as well as from the archaeology" by Chu Huaizhen.
- "Chat on Karez" by Lin Kuicheng.
- "A survey of Karez system in Xinjiang" by Qian Yun.
- "Qanats and irrigation cultures in Iran" by Michael E. Bonine.
- "Development and recent changes in the Karez (Qanat) systems of Iran" by Peter Beaumont.
- "Karez irrigation in Afghanistan" by Daniel Balland.
- "Social aspects of Karez in East Afghanistan– a traditional pattern as observed in 1960s" by Masatoshi A Konishi.
- "The Khettaras of Morroco" by Sabbayi II. Larbi.
- "Mambos–horizontal wells–in Japan" by Tanaka Kinji.
- "Evaluation on Karez water chemistry and qualities in Turpan Basin" by Song Yudong.

- "Studies on Karez utilization and transformation between surface water and ground water" by Tang Quicheng.
- "Karez, water resources in Xinjiang, China" by Yoshio Kimoto.
- "Rational exploitation and utilization of water resources in Karez distribution area" by Ma Xiping.
- "To recover the vigor of the ancient water project–Karez" by Vygur Minup.
- "Discussion on the technical problems in transforming and utilizing Karez" by Li Yuzhu.
- "The way to reform Karez– the report on the experience of drilling artesian water in the Karez" by Zhang Chun.
- "Application of scientific technique in protecting and reforming Karez in Hami" by Liu Yujie.
- "Developmental prospects of Karez in Turpan Basin of Xinjiang" by Guo Xiwan.
- "Mancheng phreatic water flow intercepting canal installation in Zhuanglang river valley of Gansu province" by Yang Xijing.
- "The water conservancy regionalization and the perspective of Karez in Turpan region" by Yang Wei.
- "Artificial recharge of Karez" by Abdul Karim Behnia.
- "Rehabilitation and reinforcement of Karez in Tocsun country" by Aihamaiti.Wupur.
- "The artificial ecologic environment and its animal utilization of the Karez" by Lan Xin.

5.16.4 International Symposium on Qanats (Yazd, Iran, May 2000)

This Conference was held in Yazd from 9th to 12th of May 2000 in the historical city of Yazd. The importance of qanats in central plateau of Iran encouraged the authorities of the Islamic Republic to put forward the proposal of the occasion to UNESCO, and it was approved by the 29th General Conference of this International Organization. Finally, this seminar was held in Yazd underwritten by the Iranian Ministry of Energy. The prominent objectives of the symposium were as follows:

1. Assessment of appropriate groundwater exploitation in arid and semi-arid regions.
2. The study of operation management to fulfill the water and civilization plans of UNESCO.

3. Carrying out a survey on different aspects of qanat technology including historic, cultural, social, technical, geographical and managerial affairs. The articles which met the goals of the symposium were selected and compiled in the book of the conference. Out of 220 articles received by the secretariat, 42 were presented in 9 sessions and eleven papers were selected as posters and from 28 foreign articles, 12 papers were approved by the scientific committee of the symposium and presented. These articles were all printed in two volumes of the seven books by the conference. The other books of this international occasion are: Bibliography of the Qanats, Qanat Glossary, The Qanat of Ghasabe in the Township of Gonabad, the English abstracts of the conference and the selected articles on the qanats. Five hundred national and international individuals attended the seminar, some of whom were prominent scholars. Observance of this symposium engendered two achievements:
    - Putting the plan of the establishment of an international center on Qanats into practice.
    - Inauguration of the water museum in Yazd and consequently creation of the Qanat Training Center.

Furthermore, some of the historic hydraulic systems of the province such as water reservoirs and ice-houses were reconstructed for the field visits of the participants.

The declaration of this symposium, which was certified by all the attendees included the following 13 articles:

1. Creation of an International Center on Qanats and Historic Hydraulic Structures for international cooperation and contributions for the execution of research, education and conveyance of qanat know-how.
2. Establishment of the National Committee of Qanats for collaborative deeds carried out by different involved sectors.
3. Creation of a qanat supervision office in the water involved ministry.
4. Reformation and amendment of the civil water regulations in order to keep further over-abstraction of groundwater at bay.
5. Documentation of the qanats and their historic and technical features as international cultural heritage, and safeguarding them under some global regulations.
6. Performance of an exemplary research project for the recognition, documentation and conveyance of qanat know-how and its sustainable exploitation through national and international means.

7. Attraction of public sector financial, technical and logistical support for qanats due to their groundwater extraction role, as well as encouraging the private sector for investment and social participation in order to renovate and preserve qanats.
8. Provision of the National Qanat Glossary and Atlas.
9. Regularization of the qanats running under urban areas.
10. Formation of cooperatives for qanat foremen and practitioners and seeking remedies for their job protection and their insurance in particular.
11. Integration of qanat traditional know-how with modern technology to facilitate the renovation trend and prosperity of these ancient hydraulic systems.
12. Implementation of projects on the artificial recharge of the aquifer in basins where there are qanats.
13. Prevention ofste of water by percolation from the tunnel along the qanat transport section.

The seminar was held under the chairmanship of the (then) deputy minister of the Ministry of Energy, Mr. Zargar, and Dr. Semsar Yazdi was the secretary.

5.16.5 Scientific Conference in Paris 2001

The subject of this conference was "Ancient Irrigation Systems: Qanats and Galleries in Iran, Egypt and Greece", which took place at the Collegé de France in Paris. The chairman was Prof. Pierre Briant. This conference was attended by historians, archeologists, and geographers. The articles presented mainly revolved around the historical and archeological features of Qanats. Except for the nice introduction by Mr. Pierre Briant at the beginning of the proceedings, there is no trace of summing up of the conference or the final statement in the proceedings.

5.16.6 Les Galleries de Captage en Europe Mèditerranèene, une approche Pluridisciplinaire Spain 2001

The International Qanat Conference of Yazd (2000 AD.) received full support of UNESCO. Following UNESCO's overwhelming endorsement of this conference, a scientific seminar entitled "Subterranean Galleries for the Conveyance of Ground Water (Qanats) in Mediterranean Europe" r in French: "Les Galleries de Captage en Europe Mēditerranēene: une approche pluridisciplinaire" was held in Casa de Velazquez in Spain (June 4–6, 2001) pursuing the following objectives:

1. Better acquaintance with the technical aspects of qanats in the North Mediterranean
2. Historic analysis of this hydraulic system with regards to the qanats and water abstraction/utilization interactions in Mediterranean countries.
3. Reconsidering the application of this know-how
4. Setting up brainstorming sessions by some scientists dealing with geography, history, linguistics, ethnology and hydrogeology concerning qanats, keeping in mind that each expert observes the issue from his own point of view.

After the inauguration of the opening ceremony by the Madrid deputy of the parliament and the managing director of the Regional Water Authority of that city, the articles were presented in Portuguese, French, Japanese and Spanish. Twenty papers were presented by lecturers from France, Portugal, Japan, Spain, America, Italy, Germany, Israel, Morocco, Iran and Corsica Island. The third day of the conference was allocated to a field visit to an ancient qanat in the township of Ocania on the outskirts of Madrid and one other. Both of the qanats had been renovated. They supplied Madrid's water in bygone eras. This conference was attended by 50 participants. The proceedings included 20 articles but the entire corpus of the papers has not been printed yet. No formal declaration was issued for this seminar but the raised materials were summed up at the concluding session by one of the French researchers, Daniel Balland:

"To open a debate on Qanats, a common language is the first need. The word "Qanat" should be defined. Different names for this hydraulic system available in different languages should be collected. Differences between Qanats, springs, artesian springs and subterranean canals for ground water conveyance should be pondered over. The second step is the compilation of a common glossary of Qanats including classified words collected from all over the globe concerning the issue. Documentation of the verbal Qanat know-how is the third step. Not to lose the opportunity, we should carry out the task before the last generation of the practitioners of advanced age pass-away and bury their own knowledge. The fourth step is the precise recognition of the other ancient hydraulic structures which are in relation with Qanats and the last step to be taken is the definition of some technical scientific terms for Qanat such as "Specific Debit" which means "liter of discharge per second along per wet-zone length"

This conference also underscored the importance of the following points:

1. Re-engaging UNESCO and other scientific societies in the studies of qanats and other underground water conveyance canals and their role in water supply
2. Coordination between the researchers of different aspects of qanats
3. More global endeavors to be made in order to preserve qanats
4. Dissemination of qanat know-how

5.16.7 Oman International Conference on Management and Development of Water Conveyance Systems (Aflaj) May, 2002

Due to the antiquity and number of qanats in Oman and the necessity felt for the preservation of these sustainable hydraulic systems, the Ministry of Regional Municipalities, Environment and Water Resources of that country sponsored this conference which was held from 17th to 20th May 2002 in Muscat. According to the available records, as well as UNESCO's comments, this event can be considered as complementary to the 1st Symposium on Qanats in Yazd, i.e. as the Second International Conference of Qanats. Some eminent political public figures from Islamic countries, such as the Omani Secretary of State for Environment and Water Resources, the Secretary of State for Energy from the United Arab Emirates, Syrian Minister of Water, the Iranian Minister of Energy and the Minister of Water and Electricity of Kuwait gave lectures at this conference. The scientific papers dealing with the following different issues were presented at four sessions:

- Session one: Importance of Qanats and some remedies for their efficiency augmentation
- Session two: Promotion of Qanats, their management and ground water security
- Session three: Analysis of the running water, the ground water and their relation with Qanats
- Session four: A survey on the legal grounds of the Qanats and irrigation

The twenty four scientific papers presented during the four sessions were referred to in the declaration of the conference. The final statement of the conference is more or less the outcome of the lectures comprising three divisions:

a. Global necessity of safeguarding water resources, water management and sustainable development and the role of the water abstraction systems including qanats.
b. The second division dealt with the symposium's points of interest like efficient qanat operation methods, statistics and data processing, the interactions between surface water and groundwater, regulations and social and economic aspects of Qanats which were all raised.
c. The third division concerned the solutions to qanat (aflaj) management deficiencies, including: systematic data collection and organizing data bases; technical collaboration regionally and internationally; drawing up long term plans for the exploitation, preservation and durability of qanats; creation of artificial recharge dams to enrich the aquifers; the need for the public participation in planning, management, exploitation, preservation; and finally compilation of regulations for the betterment of the operation of aflaj (qanats).

Only the abstracts of papers and the declaration of this occasion are available at the International Center for Qanats and Historic Hydraulic Structures in Yazd.

5.16.8 International Frontinus Conference, Luxemburg, 2003

The conference "Water Supply by Qanats – Qanats as Models for the Modern Way of Tunnel Construction" took place in Luxemburg in October 2003. The theme of this conference was the construction methods of tunneling by the qanat way, as it was used in Iran and was introduced into European territories by the Romans for the construction of underground aqueducts.

This conference took place on October 2–5, 2003 in Walferdange near Luxemburg, which has one of the most important qanats in what was Roman Gaul. This conference was set up by the Frontinus Gesellschaft e. v. Köln, which is a scientific association. The scientific committee of this conference was under the supervision of Dr. Klaus Grewe from Rheinisches Amt für Bodendenkmalpflege in Bonn. Prof. Dipl- Ing Pit Kayser, Prof. Dipl- Ing Guy Waringo, Mr. Emmanuel Sallesse were the members of the organizing committee. Twenty two articles were presented at the conference, most of which investigated qanat related hydraulic structures, tunnels and the techniques of the construction of qanats. The proceedings of this conference have been published by the Frontinus Gesellschaft e. v. Köln.

5.16.9 National Conference on Qanats, Gonabad – Iran, 2004

This conference was held from 6th to 7th of May, 2004 in Gonabad, a township in the province of Khorasan. It was underwritten by the governor's office of that city. Reassessment of qanats with regards to their role in the urban development of the area was the main objective followed by the authorities. The framework of this symposium was based on the ten following points:

1. Thorough assessment of the geographical and historical aspects of qanats,
2. Reviewing the technical and engineering aspects of creation, operation and rehabilitation of qanats,
3. A review of the cultural aspects of qanats (archeological and historic features as well as civilization, culture and literature),
4. Qanats and society (politics, law, anthropology, cooperation, social institutions, migration, permanent residence and endowment),
5. Assessment of qanat's tourist attractions,
6. Economic assessment of qanats (ownership, productivity and profitability of qanat based societies, etc.),
7. A comparison between Qanats and the other groundwater mining means,
8. A case-study of the ganats in Gonabad and their monograph,
9. A study on the Ghasabeh Qanat as the oldest Qanat enjoying the deepest mother-well in the world,
10. Gonabad Qanat and its development.

Overall 150 articles were received by the secretariat and conveyed through the scientific committee by Mr. Abdullah- Lotfi, the executive secretary of the conference. After detailed consideration, 42 articles were selected to be presented. A report concerning the establishment of the International Center on Qanats and Historic Hydraulic Structures was submitted by the Deputy Minister of the Islamic Republic's Ministry of Energy in charge of water affairs. The following final statement with 14 articles was the outcome of this seminar:

1. Showing appreciation for all the endeavors made by the ancient practitioners of qanats.
2. Public sector should pay more attention to this historic hydraulic structure.
3. Life and health insurance should be considered for qanat practitioners
4. Indigenous know-how of the qanats should be merged with modern technology.

5. The involved governmental organizations should prepare a comprehensive data bank for qanats in Iran to be accessible to enthusiasts.
6. Qanat bounds should be closely considered when permits for the creation of deep wells are to be issued.
7. Establishment of water museums in cities with qanats seems necessary.
8. Creation of mechanized irrigating systems should be put on agenda.
9. Documentation of those qanats which have not been registered yet.
10. Execution of remedies in some qanats for tourist visits.
11. Securing those qanats running beneath residences agains collapses by the use of the modern technology.
12. Launching master's degrees for qanat studies in one of the national universities in order to train the needed man power for the operation and reclamation of qanats by means of modern technology.
13. More studies should be done on the Qanat of Ghasabeh as one of the prominent galleries of the world, as well as putting in it on the international cultural heritage list of UNESCO.
14. Necessity of the establishment of a permanent secretariat at the International Qanat Center of Yazd where the outcomes of the research can be assessed annually.

An exhibition displaying the traditional techniques and facilities for groundwater abstraction was set up during this conference.

The participants visited the Qanat of Ghasabeh partially renovated for the visit by the Jihad Agriculture and the governor's office of Gonabad so that they could walk along the gallery.

Information concerning this seminar can be accessed in the governor's office of Gonabad and the International Center for Qanats of Yazd.

5.16.10 International Conference on Qanats, Kerman 2005

This conference was held on 23–24 November 2005 in Kerman, one of the arid provinces of Iran. Benefiting from numerous active qanats with considerable discharge was the reason why this city was singled out to be the host of the symposium. The Iranian Academic Center for Education, Culture and Research (ACECR) of the University of Shahid Bahonar organized this occasion in collaboration with the Regional Commission of Drainage and Irrigation and the Iranian Irrigation and Water Engineering Association. Mr. Reza Kamyab Moghaddas was the executive secretary of the seminar. Sixty one articles out of 180 received were selected and presented in Persian and English. Safeguarding qanats as cultural heritage and efficient water utilization for sustainable food production systems were the objectives followed by this conference occurrence. The presented papers revolved around the following categories:

1. A review of qanats' different aspects
2. Qanat and environment
3. Management and preservation of qanats
4. Economic aspects of qanats
5. Runoff dispersion and artificial recharge of the aquifer
6. Irrational use of qanats
7. Assessment of the structural deficiencies of qanats.
8. Mathematical and physical modeling of qanats
9. Hydrology and hydrogeology of qanats
10. Qanat drainage
11. Application of modern technology in qanats

One of the subsidiary measures to this conference was the renovation of a part of the Jupar Qanat enjoying a discharge of 150 liters per second which was visited after the conference sessions.

The received articles were compiled in one English and two Persian books. The following books were the other publications of this conference:

1. Proceedings and qualifications of the researchers
2. Bibliography of Qanats by Abdul-Karim Behnia
3. Qanat of Jupar in Kerman by Hamid Taraz
4. A Survey on the Qanats of Bam from technical and engineering points of view by Ali. A. Semsar Yazdi, Majid Labbaf Khaneiki, Behruz Dehghan Manshadi

5.16.11 International Conference on Kahriz Systems as Architectural Monument, Sustainable Source of Water and Factor of Social and Economic Development, Nakhchivan Autonomous Republic, 2009

On 24–26 September 2009, the international conference on "Kahriz Systems as Architectural Monument, Sustainable Source of Water and Factor of Social and Economic Development" was held in Nakhchivan with joint cooperation of International Organization for Migration (IOM), Nakhchivan State University and the Swiss Agency for Development and Cooperation (SDC). Scientists from Japan, Pakistan, Iran and Oman, representatives of

UN, OSCE, IOM as well as USAID and SDC, officials of the Embassies of the USA, Japan, South Korea, Switzerland, Germany and Norway in Azerbaijan, the consul-generals of Turkey and Iran in Nakhchivan, researchers of Azerbaijan National Academy of Science, academicians of the Nakhchivan State University, officials of Azerbaijani government and local government of the autonomous republic, representatives of various NGOs and students attended to the conference.

5.16.12 IWA Workshop on Traditional Qanats Technologies, Marrakech, Morocco, 2013

Between 24 and 26 October 2013, an international workshop on Traditional Qanats Technologies was organized by Morocco National Office for Electricity and Potable Water and the International Institute for Water and Sanitation in cooperation with the International Water Association. The rationale of this workshop puts a high value on the benefits and knowledge of this sustainable water system that could make an important contribution to the rational planning and water supply in the cities and territories as part of an approach toward resilience against the ongoing climate change.

The main objective of this workshop was to bring together the water management players, engineers, policymakers, sociologists, economists, archeologists and executives of water services of national and local authorities in order to:

- Safeguard the knowledge of traditional qanat systems and standardization
- Exchange the knowledge, expertise and experience
- Facilitate the maintenance of such traditional systems and to meet the specific needs in arid and semiarid areas as a remedy to the climate change impacts
- Establish a specialist IWA working group on traditional qanat systems and publish an IWA-publishing Book on "Traditional Qanats Systems"

5.16.13 International Conference on Karez on Cultural Borders, City of Bidar, India, 2017

This conference took place in the historic city of Bidar in Karnataka, India between 29 and 31 October, 2017. The conference was organized by the Indian Heritage Cities Network Foundation (IHCNF) together with the International Center on Qanats and Historic Hydraulic Structures (ICQHS) in cooperation with UNESCO New Delhi.

The conference was intended to bring together the different pieces of the existing knowledge on Indian Karez and from across the world, because he more we know about this heritage of India, the better we can learn from and preserve it specific to the Karez/qanat technology in India. This event intended to bring together all the scholars who had so far studied on this system in India but also elsewhere in the world and pool their knowledge and experiences. The scholars' contribution was considered very helpful to better identify and preserve this valuable heritage in India and also highlight its potential as a World Heritage in the future. The main objectives of the conference were to:

- Explore historical records and archeological evidence on the geographical diffusion of qanat in world and the role of nations in spreading qanats/karez
- Identify intercultural interactions regarding qanat/karez technology
- Echoing the essential message of qanat/karez on cooperation, peace and reconciliation
- Examine different aspects of cooperation and social convergence in qanat/karez
- Portray qanat/karez as a bonding string going through different nations at the grass root level

5.16.14 National Conference on Qanat; Eternal Heritage and Water, Town of Ardakan, Yazd, Iran, 2019

This conference took place on 20–21 February 2019, in Ardakan a historical town 60 kilometers northwest of Yazd, Iran. Payam Noor University organized this conference in cooperation with International Center on Qanats and Historic Hydraulic Structures. This conference was focused on the potentials of qanats and the lessons that indigenous water technologies still hold for us. The main themes of this conference were as follows:

- Qanat, its attractions and potentials for sustainable tourism
- Qanat as an example of water demand management and adaptation to environment
- Technical, artistic and innovative aspects of qanat construction, maintenance and rehabilitation
- Socio-economic dimensions of traditional water management of qanat
- Qanat on the horizon of the future

170 papers were submitted to the conference, of which 102 papers were accepted by the scientific committee. The papers were delivered as oral presentations at parallel sessions or as posters.

5.17 Conference Results in Summary

Table 5.6 shows that there were just two conferences held on qanats from 1980 to 1990 and two others between 1990 and 2000. However, the six symposiums held from 2000 to 2005 bear witness to the fact that the reconfiguration of this traditional know-how as an ancient foundation for designing sustainable production systems coexistent with nature was a global concern. As indicated in the following tables, not only the developed countries but some developing nations whose qanats play a vital role in their rural economy have so far hosted qanat symposiums. However, these occasions differed according to the objectives and conceptions, as reflected in final statements. The priorities, of the objectives, concepts and declarations have been assessed in Table 5.7 through concept analysis and examining the issues emphasized the most. The concepts have been identified and prioritized based on their importance.

In the developing countries, notably Iran, the concept of "maintenance and rehabilitation of qanats" has always been the focal point of the conferences. The focuses of such conferences have been on the concept which

Table 5.6 Interval between the conferences on qanat

Interval Between the Conferences (Year)	Year	Symposium Venue
6	1981	Mashhad (Iran)
6	1987	London (UK)
7	1993	Urumqi (China)
1	2000	Yazd (Iran)
0	2001	Paris (France)
1	2001	Madrid (Spain)
1	2002	Muscat (Oman)
1	2003	Luxemburg
1	2004	Gonabad (Iran)
4	2005	Kerman (Iran)
4	2009	Nakhchivan (Azerbaijan)
4	2013	Marrakech (Morocco)
4	2017	Bidar (India)
2	2019	Ardakan (Iran)

Table 5.7 The most outstanding concepts of the symposiums

Technical and Engineering Analysis	Legal and Managerial Analysis	Descriptive Analysis of Qanats	Historic, Archeological Analysis	Social Analysis	Qanats and Sustainable Irrigation Systems	Preservation and Reclamation of Qanats	Conference Venue
3	–	2	–	–	4	1	Mashhad (Iran)
–	–	–	1	–	–	–	London (UK)
–	–	2	1	–	–	–	Urumqi (China)
3	–	2	–	5	4	1	Yazd (Iran)
–	–	–	1	–	–	–	Paris (France)
–	–	–	1	–	–	–	Madrid (Spain)
–	4	–	–	3	2	1	Muscat (Oman)
1	–	–	2	–	–	–	Luxemburg
–	–	–	3	2	–	1	Gonabad (Iran)
–	3	–	4	2	–	1	Kerman (Iran)
4	–	–	–	–	3	2	Nakhchivan (Azerbaijan)
1	–	–	3	–	–	2	Marrakech (Morocco)
–	4	–	2	3	–	1	Bidar (India)
–	–	–	4	2	3	1	Ardakan (Iran)

can be incorporated into the sustainable production system, whereas at the conferences in the developed countries the historical and archeological aspects of qanats have usually been the center of attention. Therefore, the conferences held in these two groups of countries seem somehow different from conceptual point of view. In the first group of countries, there is still hope for the revival of qanats, which is supposed to be a suitable factor to reconstruct the sustainable production system. However, in the second group of the countries, qanats are mostly considered a subject of interest for some scientific fields, particularly history and archeology, and their attitudes toward qanats often seem the same as toward historic monuments.

References

Aflaj Inventory Project Summary Report, (2001). Ministry of Regional Municipalities Environment and Water Resources, Sultanate of Oman, p. 10.

Al-Ghafri Abdullah S. and others, (2002). Traditional methods of water distribution in aflaj irrigation systems of Oman, Oman International Conference on the Development and Management of Water Conveyance Systems (Aflaj), Ministry of Regional Municipalities Environment and Water Resources, Oman.

Al Marjeby Aley A., (2002). Development and management of water resources in the Sultanate of Oman, Oman International Conference on the Development and Management of Water Conveyance Systems (Aflaj), Ministry of Regional Municipalities Environment and Water Resources, Oman.

Al-Suleimani Zahir, (1999). Oman Country Report, Project of the Preparation of the Source Book for Alternative Technologies for Freshwater Augmentation in West Asia Region. Internal Technical Report, ACSAD, Damascus, Syria.

Bernardo López Camacho, Canal de Isabel II, Les anciennes galeries d'approvisionnement en eau daus la région de Madrid, Colloque: "Les galeries de captage en Europe Méditerranéene, 4–6 Juin 2001, Madrid.

Bernardo López Camacho, le Qanat de la Fontaine Monumentale (Fuenle Grande) d' Ocaña (Tolède), Colloque: "Les galeries de captage en Europe Méditerranéene, 4–6 Juin 2001, Madrid.

Canal de Isabel II, (1954). Bosquejo histórico y anecdótico Memoria de los cien primeros años del Canal de Isabel II. Madrid.

Cathary M., (1964). Tentative 1961–1963 dans le périmetre de Marrakech, Paris: CHEAM, pp. 68–70.

Chi-fei Liu, (1990). studies on the origin and utilization of karez in Xinjiang, International Symposium on Karez Irrigation in Arid Region, China.

Cirella Anna, Foggara Project Incomed – Ipogea Water Cultural Heritage: A Dimension of Participatory Approach in the Mediterranean, Ipogea – Italian Research Center on Local and Traditional Knowledge, p. 4.

G. De Feo, S. De Gisi, C. Malvano, D. Capolongo, S. Del Prete, M. Manco, F. Maurano, E. Tropeano, (2009). Historical, Biological and Morphological Aspects of the Roccarainola Qanat in the District of Naples, 2nd International Symposium on Water and Wastewater Technologies in Ancient Civilizations, Bari, Italy.

Goblot Henry, (1979). Les Qanats Une technique d'acquisition de L'Eau, Ecole des hautes etudes en sciences sociales centre de recherches historiques.

Joffe E.G.H., (1989). Khattara and other forms of gravity-fed irrigation in Morroco, Qanat, Kariz and Khattara: Traditional Water Systems in the Middle East and North Africa, edited by Peter Beaumont, Michael Bonine and Keith McLachlan, the Middle East Center, School of Oriental and African Studies, University of London, p. 199/201/207.

Kobori Iwao, (2009). Case Studies on Kahriz in China and Manbo in Japan, International Conference on Kahriz, Nakhchivan.

Labbaf Khaneiki Majid, (2017). Indian Version of Qanat in Bidar, Karnataka, Report of mission to India, UNESCO-ICQHS's archive, International Center on Qanats and Historic Hydraulic Structures.

Laureano Pietro and Sciortino Maurizio, (2006). Global desert outlook, United Nations Environment Program, Chapter 2; People and Desert, p. 12.

Laureano Pietro, Traditional Techniques of Water Management: A New Model for a Sustainable Town and Landscape, From the First Water Harvesting Surfaces to Paleolithic Hydraulic Labyrinths, Urban perspectives, p. 5.

Lightfoot Dale, (2009). Survey of Infiltration Karez in Northern Iraq: History and Current Status of Underground Aqueducts, A report prepared for UNESCO Iraq Office.

Maria Bascones Alvira, (2001). Irene de Bustamante Gutiérrez, Bernardo López, La funete del Berro, Colloque: "Les galeries de captage en Europe Méditerranèene, 4–6 Juin 2001, Madrid.

Megdiche Feirouz, (2018). Regards sur les paysages des sociétés hydrauliciennes du Moyen-Orient à travers les techniques des médiation identifiables à Nizwa (Oman): Aflaj et qanat, PhD dissertation, Université de Sousse.

Musayev Arzu, unpublished report, IOM Nakhchivan.

Nair V. Sankaran, (2004). Etymological Conduit to the Land of Qanat.

Nash Harriet, (2007). "Stargazing in traditional water management: a case study in northern Oman", *Proceedings of the Seminar for Arabian Studies* 37.

Roger D. Hansen, Karez (Qanats) of Turpan, China, http://www.waterhistory.org/histories/turpan/

Ruf Thierry, (2008). Summaries of a travel of water history in the world of old agricultures and organization of comparisons between hydraulic development, 2nd short course of world history of water management, UNESCO – IHE, Delft.

Sayed Sharif Shobair, (2008). National Project Coordinator and Chief Engineer, NPC and CE – FAO – EIRP.

Shouchun Wang, (1990). the study of historical geography on the origin of the karez of Turpan and its effects, International Symposium on Karez Irrigation in Arid Region, China.

Todaro Pietro, (2000). the ingruttati of the plain of Palermo, Proceedings of the First International Symposium on Qanat, Yazd, Iran, p. 47.

Weisgerber Gred, (2003). the impact of the dynamics of qanats and aflaj on oases in Oman, Internationales Frontinus-Symposium "Wasserversorgung aus Qanaten – Qanate als Vorbilder im Tunnelbau", Walferdange, Luxemburg.

Wuttmann Michel, (2001). "The Qanats of Ayn-Manâwîr, Kharga Oasis, Egypt", in Jasr, p. 1.

Xiru Zhang, (1990). The condition of formation of karez in Xinjiang and theories about their origin, International Symposium on Karez Irrigation in Arid Region, China.

6

Interviews with Local Practitioners

6.1 Introduction

Underground mining is one of the oldest jobs that humans started to do some 40,000 years ago. "Underground mining involves quite a number of both technological and cognitive pre-conditions. To begin with, it requires a preparedness to enter an alien environment which most animal species avoid, or the behavioral flexibility to manage a perhaps genetically determined cortical response pattern to fear of caves. This already provides considerable insights into the level of conscious decision making required in this context. Next, most of the underground work presupposes the availability of artificial lighting, and there is some evidence of lamps and torches having been involved in these quests. It is also obvious from several of the sites that the mining activities must have been team work, involving at least two or three people, who no doubt had to co-ordinate various aspects of their efforts. We know that a variety of mining tools were involved at that very ancient time and we can assume that items such as pointed, perhaps fire-hardened wooden wedges were prepared outside the cave. At a few sites there is evidence of the use of scaffolding, which would imply even more planning. Some of the caves are of quite difficult access, and the sheer logistics of the mining operations conducted in them must have involved engineering skills of an order of magnitude few archaeologists would be currently prepared to credit any 'pre-Upper Palaeolithic' people with" (Morton, 1996). At Lion Cave in Swaziland, ancient miners cut a tunnel 25 feet wide, 30 feet deep, and 20 feet high. This tunnel was cut into a cliff face 500 feet tall. This is apparently the oldest known mining operation. The activity has been securely dated to go back at least 43,000 years by carbon 14 and probably goes back even further to 70–110,000 years ago (Dart and Beaumont, 1971).

Tens of thousands of years of underground mining history accumulated a heritage of traditional know how which has been handed down from generation to generation. This traditional knowledge has been acquired through thousands of years of trial and error as well as through an intuitional cognition the ancient man had toward nature.

This part is just a sample of documenting the knowledge and expertise of the traditional qanat masters who are on the verge of passing away. During the last decades, the advent of new technologies broke the chain of traditional know how which was once handed down from father to son. Nowadays it is very difficult to find a qanat master whose profession has been adopted by his son. The qanat masters' offspring see this profession as menial job and prefer to lead an easy life rather than make their father's job their busyness. Therefore if we linger to document this intangible technical legacy, it would vanish soon. The following interviews we had with some prominent qanat masters in the province of Yazd encompass the most frequent questions about qanat construction and maintenance. We hope that we can accomplish this work in a separate volume in near future.

Ali Moghanni Bashiyan

This great man who has complete proficiency in Qanat issues was born in 1926 in Ardakan which is a township in the province of Yazd. He does farming & animal breeding besides the Qanat work. He has worked in the province of Fars too.

1. How do you know where to construct a qanat?

The best place is near an extensive alluvial fan on a favorable slope; a good example is near Shirkooh Mountain, which has an aquifer stretching from the foothills to the plain of Yazd-Ardakan. The upslope qanats bring water onto the cultivated lands above the second group of qanats and so on. After finding the best spot, we sink three trial shafts in a row 500–700 meters apart; these reach the water table at varying depths and this means that we can estimate the hydraulic gradient and anticipate where the horizontal tunnel and the surface will intersect.

2. What does the distance between the shafts depend on?

The distance between the shafts depends on their depth; in most cases the distance between them is twice their depth, more if a well is sunk in hard soil and rocks. For example a 30 meters well sunk in rock may be 70 or 80 meters away from the next well.

3. What is the diameter of a well?

A normal well is 60 cm in diameter, but a very deep well could be 90 cm to provide ventilation. It was common to use a pick axe with a 30 cm handle to measure out the diameter of the well, rotating it on its pick like a compass to draft the circumference before digging (Figure 6.1).

4. How do you dig shafts in a gallery running below the groundwater level?

We dig downwards to prevent water from seeping into the well and accumulating. By digging overhead we ensure that water flows along the tunnel to the surface as soon as it seeps (Figure 6.2).

Figure 6.1 Using a pick axe to measure out the diameter of the well, rotating it on its pick like a compass to draft the circumference before digging.

Figure 6.2 Groundwater naturally seeps into the shaft well up to the water table.

Figure 6.3 Water level should be just knee high, otherwise there is an obstruction somewhere in the tunnel.

5. How do you keep the gradient of the tunnel constant?

The water level should always be up to our knee and when it goes above we know that the gradient has changed. The width of the tunnel should be about 60 cm and the height should be 1.5 to 1.75 m high (Figure 6.3).

6. How do you construct qanats that run on two different levels and intersect each other?

No qanat should run under the water production section of another qanat otherwise the lower one would drain the available groundwater from the

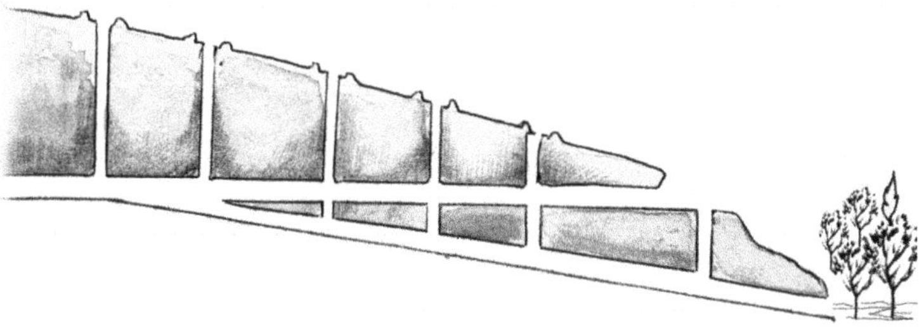

Figure 6.4 Digging another tunnel under and parallel to the old one, known as "tahsoo-bonsoo".

upper one. To determine whether a qanat cuts under the water production section we gouge a hole in the wall of the gallery to see if there is any seepage in it. If there is none we allow the lower qanat to keep making its way upslope. Sometimes we connect the two qanats with a well called a "gozare".

7. What is "tahsoo-bonsoo"?

Sometimes the groundwater level sinks so fast that the qanat practitioners have no opportunity to extend the tunnel but have to deepen it. If the water table declines so far that even deepening the tunnel no longer works, they start to dig another tunnel under and parallel to the old one. They use the shafts of the old qanat to do this and the practice is known as "tahsoo-bonsoo" (Figure 6.4).

8. What is "raff"?

"Raff" is a small niche dug into the wall of a tunnel to put our stuff in.

9. What do you do when you come across a dangerous creature like a snake or scorpion, or even poisonous gas?

Snakes are very common and we kill them if we find any. Apart from snakes there are few creatures in the active qanats. If we come across gas, we dig two wells five meters apart and joined by short tunnels. These are called "jofte-badoo" and they facilitate the circulation of fresh air in the tunnel (Figure 6.5). They are not normally needed in the water production section.

10. How do you protect a qanat from the flash floods?

If a shaft has been sunk in the middle of a valley that is subject to seasonal flash floods, we have to protect it. First we line the wells with brick, stone and cement from the bottom to 12 meters below the surface, where a concrete lid

Figure 6.5 Twin wells to facilitate the circulation of fresh air in the tunnel.

is installed. From here to the surface the well is filled up with cement, lime, clay or brick to prevent surface runoffs from leaking into the well. We always line the shafts to prevent them from crumbling.

11. How do you reduce percolation in the water transport section?

In sandy soils, we mix soft clay with water and knead this mixture before spreading it on the floor and trampling it to fill up and seal any tiny cracks. We call this practice "koor kardan-e cheshm-e zamin" – 'blinding the earth's eyes'. If there are many cracks we also use ceramic or concrete hoops. Most of the qanat masters avoid doing this laborious business unless they have no choice.

12. What is "ab-harzi"?

"Ab-harzi" means checking the tunnel to see if it is clear. It should be done once or twice a year but the intervals between "ab-harzis" depend on the type of ground the qanat cuts through as well as its length. To undertake this workers walk the tunnel from its outlet to its mother well. They take turns climbing down the qanat and walking along its tunnel until the job is finished.

13. How can you save qanat water in winter when it is not needed for irrigation?

In our region water is never left useless and it is not true that the qanat water is wasted in the winter. Instead it is used to fill reservoirs or irrigate particular crops. For example the qanat of Sadr Abad is used to fill the reservoirs of

Ardakan and the villages Tork Abad and Ahmad Abad while the water of the qanat of Nosrat Abad irrigates pistachio orchards. We use "goorabs", or traditional dams, to harvest seasonal runoffs and help them recharge aquifers.

14. How can we revive qanats which have been abandoned?

Unfortunately, there is no hope for such qanats unless all the pumped wells surrounding them are shut down. There are also very few places to build new qanats as all the good places have been used.

15. Which is more effective, extending or cleaning a qanat?

That depends on the condition of the qanat, if it is in soft soil and therefore prone to collapse, then dredging is better.

17. What is a "mirab"?

The "mirab" is in charge of dividing the water among the farmers, an incredibly important position in arid land communities, so he must be of good character, honest and reliable. The farmers allocate a portion of their water shares to cover the wages of the "mirab" and his assistants – for example the "mirab" may claim an hour of irrigation time per week from a farmer in return for his services. The money that is collectively used to finance the qanat, which is called "nafaghe" is separate from the "mirab's" fees.

18. What is "maghsam"?

A "maghsam" is an outlet built across a ditch to divide water which is used where the output of a qanat is considerable and two or more villages are entitled to water.

19. What if somebody does not want your qanat to run across his land?

In the past it was impossible to stand up to powerful landlords but after land reform, when lands were redistributed among the peasants, a qanat could run across the lands of several small owners. Therefore it has become necessary to negotiate with several people while extending a qanat. If we don't succeed with everyone we have to file a complaint; in most cases the law is on our side, because every qanat has a bound in which nobody has the right to cultivate or build anything. The court may even force the owner or owners to sell us this land for a fair price. If a qanat is very deep it is not necessary to envisage a bound for the gallery. But we must still be careful not to damage the shafts by cultivating or building above them.

Hasan Karami

This hard working practitioner was born in 1908 in Majumerd on the outskirts of Ashkezar, a township in the province of Yazd. He has worked in Qanats since he was 9 years old. His poetic nature attracts everybody. He has worked in Siatan & Bluchestan & Khorasan for many years.

1. Why do qanats dry up?

Ancient qanats from Zoroastrian times dried up naturally as the groundwater was depleted (Figure 6.6) or because they were abandoned. Now the main reason is pumped wells. These should not be a problem if they are drilled far from the qanats but if they are within the boundary, they are very bad. In fact, if we are really concerned about qanats we should fill in the wells within the boundaries of qanats and wait for groundwater to filter into the qanats. But this process may take several years and we need to work on the qanats now. One of the most significant events as far as qanats are concerned was the land reform after the Revolution in 1979. This led to the annihilation of qanats as landlords moved to the cities and no longer cared about the

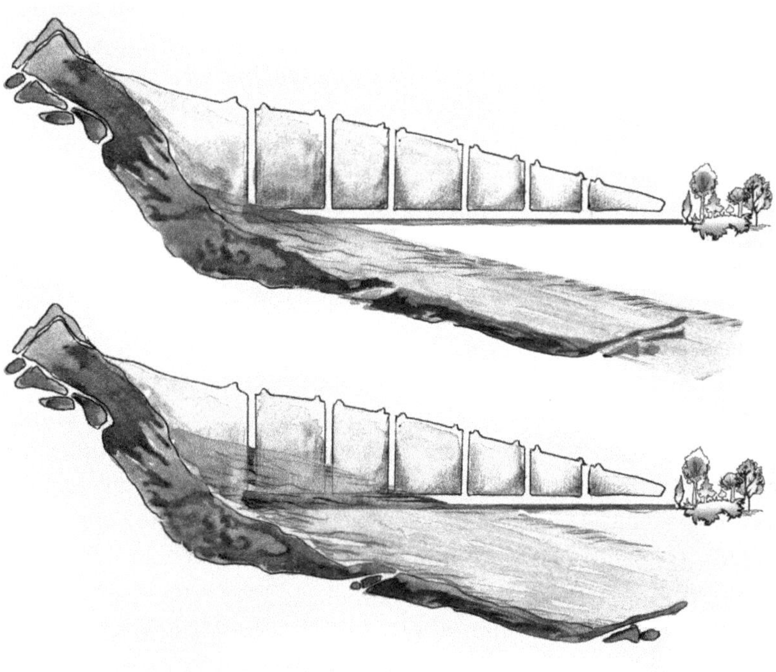

Figure 6.6 Position of the water table and its impact on the infiltration area of the qanat.

qanats. The peasants who took over feudal lands could not afford to pay for the maintenance of the qanats, so many were abandoned.

2. How can we preserve the qanats that are still active, for example mountainous qanats?

First of all we should not drill any more wells nearby. For example in Ali Abad there are nine qanats which are still active because there are no wells nearby. In the past qanats were excavated so skillfully that they did not affect existing ones nearby. All qanats ran in the same direction and at the right level so that hundreds of qanats could exist side by side without damaging each other. Qanats that have become dry because of wells cannot be saved but those that still have water in them can be extended as this helps to prolong their productivity.

3. Would it have been better to build 50 qanats than drill 100 wells in Yazd ?

No, there is no room here for new qanats and the water table has dropped so much that we cannot reach the groundwater even at 100 meters. All we can do is to extend the existing qanats to increase their discharge, although the water they provide can no longer meet the needs of the growing population.

4. Is it true that the qanat of Ebni is the oldest in Yazd ?

Yes. This qanat has rectangular wells, which shows that it is very old. We call such wells Zoroastrian wells because at the time that they were dug Iranians were Zoroastrian. Rectangular wells make it easier to find the right direction underground without navigation tools but they create more debris so after navigation tools were invented, people stopped digging rectangular wells. There are also rectangular wells in Kasnaviye and Nosratabad.

5. What is a 'rassi' and how is it used?

A 'rassi' is a wooden stick with two stones suspended from each end. The stick is placed over the mouth of the well pointing in the direction of the gallery and the stones are allowed to fall down the well so that the ropes form a plumb line. To keep the tunnel straight we put a light behind us and we turn back once in a while to check. If the light fades we are going in the wrong direction.

6. What is the first thing which should be done to construct a qanat?

First we dig two wells of the same depth with the bottoms parallel to the horizon. If the water seeping into the wells is of the same level, we have to dig a very long tunnel to bring the water to the surface, and we usually don't bother unless the qanat owners don't care about money.

7. What is a 'goorab'?

A 'goorab' is an earthen dam for harvesting water to recharge an aquifer. The word consists of 'goor' which means grave, and 'ab' which means water. A 'goorab' should be built above the qanat's mother well. During the rainy season, the runoff builds up behind the 'goorab' and gradually seeps into the ground. Sometimes, a 'goorab' is more useful to a qanat than extending the tunnel. For example, there are two 'goorab' above the qanat of Sadigh Abad that constantly drain an eight liter per second, although its tunnel is never extended.

8. Is it true that the gradient of the tunnel should be 2:1000?

The gradient can be 1:1000. If the tunnel is too steep, the tunnel may go over the water bearing zone without meeting groundwater and even if it does the water would flow so fast that it would erode the tunnel.

9. How can you prevent winter water being wasted?

Where the qanat cuts through hard soil, it is possible to build an underground dam – such as the one in Shahr-e Babak. This is a brick wall built across the tunnel where the water production section ends and the water transport section begins (Figure 6.7). There are four outlets sealed with bricks and when the farmers need water they just pull the bricks out. This technique cannot be used in qanats running through soft soil, because the build-up of

Figure 6.7 Brick wall built across the tunnel to dam water in winter when water is no longer needed.

water can damage the structure. To seal the floor of the qanat we spread a layer of soft silt on the floor of a tunnel and trample it under our feet so that the mud filled every crack. We try not to run a qanat across a river bed as the wells can be damaged by flash floods, but if there is no alternative then we seal the mouth of any well with a concrete lid and lime cement.

10. How do you know if a place is suitable for qanat construction?

First we check if there are any foothills to provide an appropriate gradient, for it is not possible to dig a qanat on flat ground. We can also see from the vegetation if there is a good supply of groundwater and its quality. However, the best indicator is a borehole. There is a poet who wrote that, "the value of hearing is never the same as that of seeing" and we can hear the underground water in a good spot. When we have found the right spot we dig to about 50 cm below the water level.

11. How do you designate the boundary of a qanat?

There is no law determining the bound of a qanat as it depends on the type of soil it runs through and other factors such as its depth. For example if a qanat is in hard rock, we can dig another qanat just 200 meters away.

12. How far apart do you dig the wells?

In most cases, the distance between the wells is equal to their depth. For example if a well is 50 meters deep it is 50 meters away from the next well. The deepest well I have ever seen was 80 meters deep by the qanat of Sadr-Abad. The maximum distance between the wells can be twice their depth. If the wells are far away from each other, we have to dig the twin wells in order to ventilate the tunnel. The well is at least 78 cm in diameter to allow room for the well digger. He will use a short-handled pick for this work and will climb up and down the well using niches dug into the side of the well, forming a kind of ladder, known as a 'paraf' (Figure 6.8).

13. How many people are on a work team?

There are always at least three people but this can go up to eight if the excavation of the tunnel and the wells is done at the same time: two people on the pulley, two for moving the debris, two for digging the tunnel and two for digging the wells. If the rock is hard, there are more people digging and fewer removing debris.

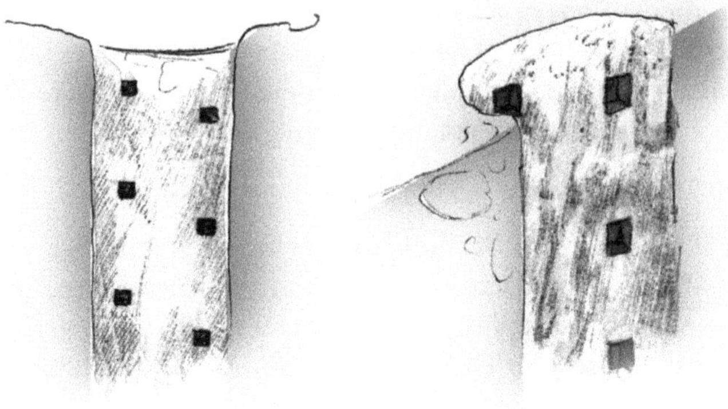

Figure 6.8 Niches dug into the side of the well, forming a kind of ladder, known as a 'paraf'.

14. What is the difference between the water transport and the water production sections of a qanat?

There is no sediment on the tunnel walls of the water production section because of the seepage, but there is in the water transport section of the qanat. The length of the two sections depends on the type of soil through which the qanat is being dug; the softer the soil, the shorter the water production section.

15. What types of pick do you use to excavate a qanat?

We use L-shaped picks with sharp curved points and a wooden handle and these vary slightly with the kind of digging that we will be doing. We use a pick that weighs 2–3 kg and has a 40–50 cm handle to dig the end of tunnel. For dredging and cleaning up the bottom of the tunnel, another pick is used that weighs 3–6 kg and has a 90–100 cm handle.

16. What about the buckets?

In the past buckets were made of goatskin not of cow leather, because leather was so expensive. Nowadays we use rubber. A normal bucket is 25 cm wide and 60 cm deep. In order to keep the mouth of the bucket open we put a stick on rim, turn the rim back over the stick and sew it, so that the stick forms a ring. When we used leather the hides were soaked in a mixture of water and lime, before being tanned and then soaked in water and pomegranate skins. It cost us 0.1 cent for five tanned goatskins but the buckets tore every 10 or 20 days, so we often had to buy new ones.

17. And the lamps?

Nowadays we use carbide lamps, but 60 years ago we used lamps fueled by castor, cotton seed or arugula oil. Animal fat was too expensive and produced more smoke. We would place a spare lamp in the tunnel on a special shelf in the wall of the gallery.

18. What kind of rope do you use?

The best one is 'sazoo' a very strong rope made of palm fiber – a 20 meter long sazoo costs about 0.5 cents a meter. In the deeper wells we tie two or more ropes to each other to reach the bottom. We do not use cotton because it is not so strong and absorbs too much water.

19. What is meant by 'saboo'?

A 'saboo' is a small copper bowl with a tiny hole in the bottom that we float in a larger bowl of water. Water starts to fill the 'saboo' and the time that it takes to sink is a unit of measurement for irrigation time. The time it takes to sink the 'saboo' varies from place to place but in our region it is 11 minutes. The person who regulates the 'saboo' filling is known as the 'saboo-kesh' and the person who oversees him is the 'mirab'. It does not matter if the 'saboo-kesh' is illiterate as his job is to count how many times the bowl sinks in the water and record this on an abacus.

20. Why is the ceiling of the tunnel arched and how do you line the tunnel with hoops?

The arch means that we have to remove less debris. We line the gallery with ceramic hoops by fixing them to the walls and stuffing the gap between the hoop and wall with stones and rubble (Figure 6.9). Sometimes, we have to chip sediment off the hoops with a small pickaxe. The hoops help strengthen the walls and ceiling of the tunnel and they are made of baked clay or concrete.

21. How many hours a day did qanat workers used to work?

Those who were in charge of removing the debris and working with the windlass worked for 12 hours a day, even eating below ground. But the qanat master who dug the tunnel worked for six hours. The main reason that workers went up in the middle of their work was to pray. We used to eat barley bread and sometimes wheat bread. Sometimes the people on the surface put a loaf of bread in the bucket and this was all we had for lunch. When we were working far from town we would build a camp underground called a 'bookan' and stay there until the work was finished.

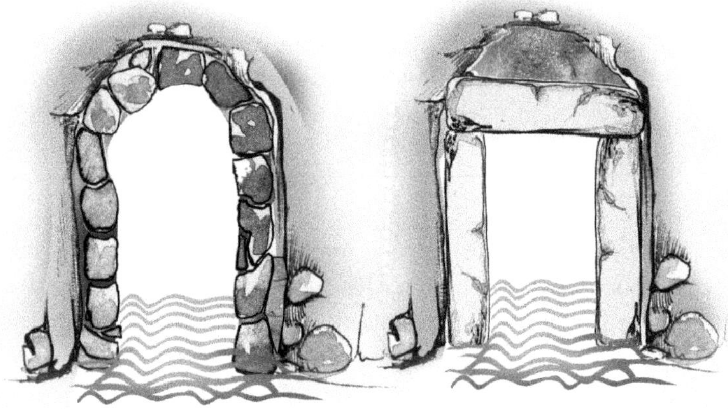

Figure 6.9 Two types of lining of gallery with stones.

22. What kind of clothes did the workers used to wear?

We used to wear simple white clothes made of canvas. It looked like a shroud and we used to say that as our work was so dangerous we were prepared for death!

23. Have you ever had an accident at work?

About 75 years ago, when I was working in the qanat of Haji Salih as a 'gelkesh' (somebody in charge of hauling debris), part of the tunnel collapsed on top of me. I was lucky that my head remained out of the rubble. Despite the terrible pressure my body was under, I could still pull myself out. Another time I remember I was being pulled up the well and just as I grabbed the rope in the middle of the well I sensed something heavy rush past me almost touching my face. It was dark and I could not see anything but a few seconds later I heard a terrible thud from the bottom of the well. Later I learned that someone had knocked a huge boulder down the well by accident. Another incident was in the qanat of Ya'qoobi. I was told that somewhere in the tunnel the surface runoff was penetrating the qanat so I had to go down the well to fix the problem. When I was walking along the tunnel in search of the crack I sensed the ground shaking under my feet. I ran back and got out of the qanat as soon as possible. What I saw on the surface shocked me. There was an incredible flood rushing over the same part of tunnel that I had been in and moments after I had escaped to the surface, the place where I had been walking was swept away. Thank God, I was not there then.

Ahmad Zahedi

Mr. Zahedi who is a veteran Qanat expert was born in 1936 in Shamsi on the outskirts of Meybod. He apprenticed this career to his father some 50 years ago. He has worked in numerous Qanats in the province of Khorasan too. Ten people are practitioners in his household now.

1. How do you decide where to construct the qanat?

Broad alluvial valleys with thick sedimentary layers and solid bedrock are the best sites for qanats. We analyze the soil, as this affects the quality of the groundwater, and then sink the first well in the middle of valley where the two alluvial slopes meet. Then we wait until the groundwater has seeped into it. We monitor the seepage rate to make sure that the permeability of the surrounding layers is adequate to support a qanat, and if it is, we start to dig the next well and the horizontal tunnel. If the permeability is not adequate we abandon that well and sink another one elsewhere.

Alluvial plains are also good qanat sites and we calculate the site for the first well by assessing the runoff from the foothills onto the flat land below. If we have to sink the well in a mountainous area, it has to be below 1500 m so that we can channel the water to the surface against the steep gradient.

It is important to determine the boundary ('harim') of each qanat in a certain area to ensure that one does not drain water from another. In mountainous areas upstream qanats have a negative impact on downstream ones, and this is also affected by the type of soil surrounding the qanat. The more permeable the soil the larger the restriction area needs to be. The opposite is true on the plains where it is the downstream qanats that affect the discharge of those upstream.

Another boundary issue occurs when a qanats run across farmland as those less than 10 m below the surface are vulnerable to irrigation and runoff. If the qanat runs through sand and loam, there should be a margin of at least six meters from the tunnel to both sides, this can be reduced to four meters if the qanat runs through clay and heavy soil. If the qanat is more than 10 m below the surface a margin of between six and 8.5 m in circumference should be left round the mouths of the vertical well shafts.

2. What happens when a shaft has to be sunk on someone else's land?

The landowner's consent is required or we can arrange to buy the land in question. Otherwise we have to bypass the land or, if this is impossible and the landowner still refuses to compromise, we can appeal to a court. If a qanat under a house or a river, we have to reinforce that part of tunnel with concrete.

Figure 6.10 Rassi: a traditional tool to find out the right direction underground.

3. How do you calculate the length of a qanat and the distance between the shafts?

The length depends on the surface gradient – the steeper the slope, the shorter the gallery of the qanat. The distance between the shafts depends on the type of soil and the depth of the shafts. The greater the permeability of the surrounding soil the closer together the shafts should be.

4. How do you dig the gallery so straight and true that it intersects the other shaft well at the bottom?

We used to use a special tool called a 'rassi' (Figure 6.10). This is a wooden stick with two stones suspended from each end on two pieces of rope. The stick is placed over the mouth of the well pointing in the direction of the gallery and the stones are allowed to fall down the well so that the ropes form a plumb line. Nowadays, we use a compass instead although we have to be careful to keep iron tools like pick axes and shovels away from the compass!

5. Could you describe the branches of a qanat and what factors affect their length?

A side branch ('dastak') serves to increase the discharge of the qanat where necessary. Side branches vary in length from 10 to 200 meters and the longer ones also have vertical shafts. Most qanats have two side branches running in opposite directions, and no additional branches are dug while existing branches are in use.

6. How wide are the wells?

Wells are usually between 60 and 70 cm in diameter, less in soft, porous soil slightly more (75 cm) in hard soil.

7. How high is the gallery?

The average gallery is 1.25–1.4 m high, but this depends on the tools we use. If we use a pickaxe, the water transport section of the gallery is 1.2–1.25 m high and 60 cm wide and the water production section is 1.4 m high and 60–75 cm wide. But if we use compressor drills, the ceiling may reach 1.60 m (Figure 6.11).

8. How long is the water production section of a qanat?

This depends on the earth surrounding the qanat and the gradient, which is usually 0.9:1000 m, but higher in mountainous areas; this section is usually

Figure 6.11 The width of the gallery does not usually exceed the distance between the two elbows.

1 km on the plains, but it rarely exceeds 200–400 m in mountainous areas. The length also depends on the depth of aquifer – the deeper the aquifer, the longer the water production section.

9. What does 'tahsoo roosoo' mean?

'Tahsoo' is the name given to a new tunnel that is dug underneath an existing one, which we call 'roosoo', when the aquifer is so depleted that the water level sinks and threatens the integrity of the qanat. When this happens we first deepen the floor of the 'roosoo' tunnel to ensure groundwater flow, but if we have to dig more than 1.5 m we abandon the 'roosoo' and dig a parallel 'tahsoo'.

10. What do you do when the tunnel is blocked by geological formations?

We dig round it unless it is so big that this is impossible, whereupon we have to abandon the planned route and dig another one.

11. What about poisonous gases in the wells or gallery?

We dig two parallel shafts 1–1.5 m apart and which are connected by small tunnels at several points to ease the circulation of air.

12. How do you stop the qanats from flooding?

First we line the wells with brick, stone and cement, and then cover the mouth with a concrete lid. In mountainous areas, where the wells could be sunk into the bank of a seasonal river, we would also build a floodwall along the river. Another solution is to block every other well at a depth of two meters.

13. How do you prevent the shafts from collapsing?

We line the shaft from the bottom to the top with stone, brick and cement and install metal rings inside and along the shaft. We also drive a round metal mould into the well, and then we fill the space between the mould and the well with reinforced concrete and iron rods.

Sometimes the gallery roof is weak and even caves in so we install concrete scaffolding rings to support the tunnel or, if the risk of a cave in is not too great, we line the sides and ceiling of the tunnel with pieces of rock.

14. How do you waterproof the tunnel?

If the tunnel is shallow, we remove the ceiling of the tunnel, line the floor with cement and build two walls along both sides. A new method is to use bentonite, (25kg:100m) and sometimes even plastic sheets, but these methods can only be used in tunnels which are not eroded. Sometimes an earthquake may cause cracks in the tunnel through which water can escape. When this happens we fill the cracks with very soft sand or clay.

15. How often do you dredge the tunnel?

Usually once every one or two years, unless the qanat runs through hard rock, when it's just once every 10 or even 20 years. The higher the density of resolved minerals in the water, the less often we need to dredge because the minerals form deposits that line the sides of the tunnel. We usually dredge the tunnel in the winter, when demand for water is low.

16. What does 'ab harzi' mean?

'Ab harzi' means walking along a tunnel to make sure that there is no obstruction and we do it twice a year.

17. What is meant by the 'irrigation cycle'?

Each farmer who has a right to qanat water possesses a particular share of this water which is determined on the basis of time. For example one farmer may be entitled to irrigate using qanat water for one hour and another for two hours and so on. The time taken for every eligible farmer to receive their quota of irrigation water is known as the 'irrigation cycle' and it lasts from six to nine days in porous soils and between 12 and 16 days in heavy soils.

The unit of time is called 'jorreh' in Yazd region. This unit is known by other names as well, such as 'saboo', in other regions. The 'jorreh' is 11 minutes, so 24 hours is equal to 128 'jorrehs'. An irrigation cycle lasting 12 days is made up of 1,536 'jorrehs' and one 'sahm' is equal to 12 hours or 64 'jorrehs'.

18. What is the main reason for the deterioration of qanats?

Qanats deteriorate because they are not managed properly and the shareholders do not care about rehabilitation and maintenance. But the drilling of deep wells is also having a very negative impact on qanats, mainly because they are being drilled within the qanat boundaries.

If a deep well has been sunk too near a qanat, the only thing that can be done is to fill the well up, but if it lies outside the boundary area of the qanat we can only monitor to ensure that it does not over-pump groundwater or else deepen the tunnel and wells or extend the tunnel through the aquifer. When planning to sink a well, you need to remember that a qanat has to be extended every year. The well should be sunk downstream from the water production section of the qanat about 1500 meters away from the structure.

19. How do you control the flow of water in a qanat in winter?

If the tunnel has been dug in hard soil and is not at risk of collapse, we install a tap where the water production section ends and the transport section starts. In winter, when demand for water is lower, we turn off the tap to save the groundwater. If the tunnel is likely to cave in, we reinforce it first by lining it.

20. How can you recharge aquifers?

We build an earth dam or dike to collect the seasonal runoff to recharge aquifers. In mountainous areas, we build the dam about 150–200 meters upstream from the deepest and last well of the qanat so that the water trapped behind the dam will seep into the earth and increase the discharge. The dam must be far enough away from the qanat so that it does not affect its structural integrity. The softer the soil, the greater the distance between the two structures. We also use abandoned qanats to recharge the aquifer by blocking the outlet and directing the surface run-off into it through one of its shafts. The water seeps into the aquifer and replenishes the active qanats nearby.

21. Can you regulate the discharge rate or does this only depend on precipitation?

When it comes to deep qanats on the plains, we regulate the discharge by extending the gallery. But the rate of discharge of most other qanats depends on precipitation; there is a fluctuation rate of 10–20 percent in qanats on the plains, and 60–80 percent in qanats in mountainous areas.

22. Given that qanat engineering is difficult and dangerous, how can we encourage people to stay in and even take up the profession?

The government should provide incentives such as special insurance and pension schemes to qanat engineers, and there should be an annual ceremony in recognition of the profession. Initiatives like this will encourage people to stay in the profession and attract young people to the job.

23. Which is more important, dredging or extending the tunnel?

That depends on the qanat. We would first check the tunnel for cave ins or serious erosion and if we found signs of these we would dredge rather than extend the tunnel. Otherwise we would extend the tunnel to increase discharge but sometimes we do both.

24. Nowadays is it worth maintaining qanats?

Absolutely! Qanats are immensely valuable economically and technically important. Demand for water is growing and qanats provide water in an environmentally friendly and sustainable way. But today, because deep wells have depleted the water table, building a new qanat is almost impossible. The only option we have is to rehabilitate existing qanats through artificial recharge projects and to ban over-pumping.

25. How many people make up a qanat construction team?

A qanat construction team is between three and six people, depending on the length of the qanat and the depth of the wells. There are four distinct categories of qanat engineer: the 'karshena', who decides where the qanat shall be dug and supervises the entire project; the 'ostad kar' or master worker, who is in charge of digging; the 'gelband', who collects the debris, and the 'charkh kesh' who operates the pulley to haul the bucket up the well. If the well shafts are far apart, another person, the 'lashe kesh' is also added to the team to drag the bucket of debris along the tunnel to the nearest shaft and to tie the bucket to the pulley rope.

26. What tools do you use?

We use a pickaxe, pulley, rope, shovel, oil or carbide lamp, electric fan, compass and plumb line to dig the qanat and we use the pulley and well wheel to haul the debris from the tunnel. There are different types of pickaxe depending on the use. To dig at the end of the tunnel use a pick weighing 2–3 kg with a handle 40–50 cm long; for dredging the tunnel we use a pick weighs up to 6 kg with a handle up to a meter long. The type of pickaxe also depends on the soil, for hard soils we use one which has a short, thick point, rather like the beak of a sparrow and we use a pick with a long, thin point in soft soils. The handles are about 40–50 cm long if they are to be used for sinking a shaft, and 50–60 cm if they are to be used for digging a tunnel, and 90–100 cm if they are to be used for removing obstructions from the bottom of a tunnel. The leveling tools we use are two wooden poles 1.5 m long with a square base and a hole in the middle with a plumb line in it (Figures 6.12 and 6.13). The steeper the gradient, the shorter the distance between the two poles.

27. What about the pulley?

In general, the diameter of the pulley is around four meters, but that depends on the depth of the shaft. The deeper the shaft, the greater the diameter

Figure 6.12 Traditional structure of a leveling tool.

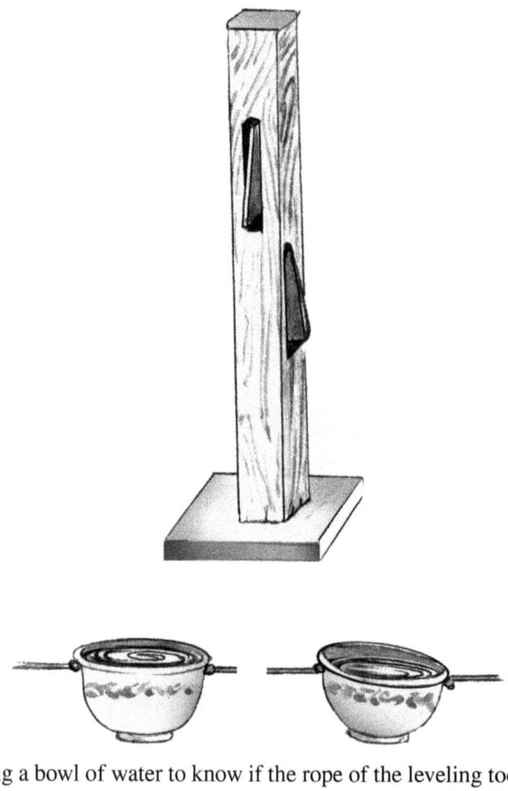

Figure 6.13 Using a bowl of water to know if the rope of the leveling tool is just horizontal.

of the pulley (Figures 6.14 and 6.15). We reinforce the legs of the pulley by digging holes on both sides of the mouth of shaft and putting the legs into the holes and filling them up with stone and dirt. We use cotton, palm fiber, steel wire and plastic for the rope – obviously the deeper the shaft, the thicker the rope and the stronger the material that it's made of. We use steel wire for the electric pulleys. In the past, we used a kind of bucket of tanned skin of sheep. But now it has been abolished and substituted with rubber bucket (Figure 6.16).

28. How do you light the tunnel?

In the past the only way to light the tunnel was to use a lamp like a small bowl full of vegetable oil in which there was a wick. But now we use a carbide lamp or electricity.

29. What facilities do you have to insure the welfare of workers?

There is a tiny chamber built into the side of the shaft about two meters below the surface where the qanat engineers can rest or change their clothes. There is a similar space at the bottom of deep shafts to provide shelter from debris falling from the bucket as it is being pulled up. In this regard, we also hold

Figure 6.14 Pulley with two buckets hauling the debris alternatively. It is easier and quicker to use this kind of pulley.

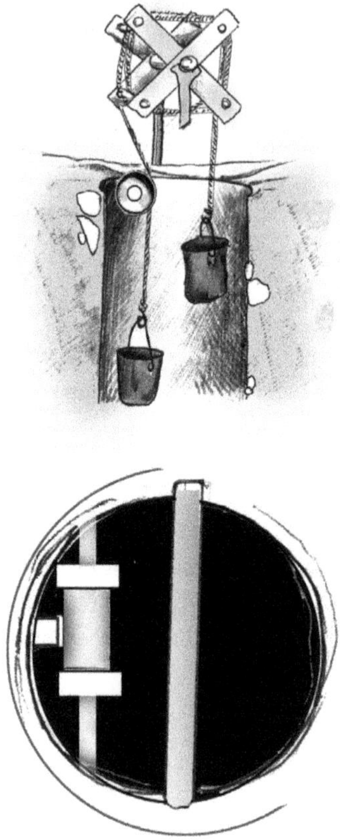

Figure 6.15 Double pulley plan.

a round wooden plate above our heads like an umbrella to protect ourselves (Figure 6.17).

30. How do you measure the volume of discharge?

We level the bottom of a ditch so that its gradient is zero and fix a wooden frame that channels the water through it. We calculate the volume of water passing through the frame and from that we can estimate the discharge rate. Some experts can also visually estimate the volume of water although this is not always accurate.

31. What does a 'mirab' do?

The 'mirab' is responsible for managing access to qanat water and for collecting the dues from shareholders in a qanat which are then used to pay

Figure 6.16 Traditional bucket and the way it is hooked.

Figure 6.17 Round wooden plate like an umbrella to protect worker against the stones falling down.

for maintenance of the structure. He is a respected member of the community who is elected by all shareholders.

32. What about the 'abyar'?

The 'abyar' oversees the irrigation of crops and makes sure that water is allocated fairly. Shareholders may come to an understanding about hiring an 'abyar', sometimes more if the qanat is large, or they may decide to manage this themselves.

33. How do you calculate the wages of people who manage qanats?

A group of farmers who irrigate their lands during a twelve hour period have a head named a 'sartaq', who is not paid cash but who can receive excess water remaining at the end of a 24 hour period. But the 'mirab' can receive a wage derived from a percentage (5–10 percent) of water shares and if he is assigned to collect money for the renovation of the qanat, he can keep five percent of the total collected. Qanat expert gets paid according to the time he spends checking the tunnels and other engineers are paid for their work.

34. What happens if a shareholder refuses to contribute to repairs to a qanat?

Everyone has contributed a sum proportionate to their shares in the qanat and someone who refuses to do so may risk having some of his shares confiscated. If a shareowner has rented out his share, we charge the tenant instead.

35. What is 'maqsam'?

The word 'maqsam' refers to the place where water is distributed. If the qanat has enough water it would all be allocated as normal but if there is a water shortage we collect the discharge in the pool before distributing water to the shareholders. If the output of the qanat is such that it can be divided among several shareholders at once, we build a 'maqsam' across the flow.

36. What are the terms used to measure the discharge of a qanat?

We measure water flow by liters per second; one 'qafiz' equals nine liters per second; one 'charkh' equals seven 'qafiz'; one 'kar' or 'jow' equals 20–25 'qafiz', and this is enough to irrigate directly with no need for a 'maqsam'.

37. Can we use abandoned qanats to dispose of wastewater?

Yes, but with the permission of its all owners. Then we block its outlet and direct the runoffs into its gallery. This procedure can help preserve our groundwater.

Mohammad Ali Fayyaz

This deceased practitioner who was about 100 years old in 2001 is one of the most proverbial well diggers of the province. He inherited this profession from his father & was very eager to convey his verbal knowledge through others.

1. How do you decide where to construct the qanat?

We look at the distance of the site from the mountains and the rivers, focusing on deep valleys. Our decision is based on gut feeling and experience, and when we have chosen the spot we dig a borehole to make sure that the water is fresh. We also study water surrounding the area and if it is fresh and good it is likely to be good in the qanat as well.

2. How long is the water transport section of a qanat?

The length of water transport section ('khoshke kar') depends on the surface slope; if the gradient is steep, the water transport section will be relatively short. The same applies to the water production section ('tare kar'), but in reverse.

3. How do you know if a borehole has been sunk into a good reservoir of water?

We first dig 50 cm below the ground and then study the mud, as an experienced qanat engineer can tell from this whether there is groundwater or not. The soil also tells us how deep to dig the qanat; if it is soft and porous, we dig a steeper gallery so that the water flows faster with less seepage.

4. How do you calculate the length of a qanat and its outlet?

The length of the qanat from the first well to the outlet is based on the surface slope. To calculate the gradient we use a tool consisting of two posts with a rope between them.

5. What does the distance between the wells depend on?

It depends on the depth of the wells; a well 50 m deep should be 100 m away from the next well no matter whether it has been sunk in the water transport or water production section. A qanat master engineer may decide to change this distance; if the distance between the wells exceeds 100 m, we have to make the ceiling of the tunnel high enough for two workers to drag the debris along the tunnel.

6. How do you find the right direction while digging a tunnel?

Nowadays we use a compass, but in the past we used a plumb line made of a stick with two stones on ropes tied to each end. We placed the stick in the mouth of the well so at the bottom the stones would be orientated toward the right direction. The person in the tunnel held a light just behind one of the stones so that there was only one shadow on the opposite wall, and this pointed the workers in the right direction.

7. How wide should a well be and how do you climb down and up it?

A well is usually 70–80 cm in diameter, so that the worker can wield the pickaxe. We climb up and down the well using a rope or steps dug into the side of the well.

8. How high should a tunnel be?

The tunnel is about a meter high and between 60 and 70 cm wide. If the worker wants to speed up digging, he will dig a tunnel 60 cm wide although it would be more difficult for him to pass through.

9. If a tunnel runs through a water bearing zone, how do you dig the well?

We use a method of digging called 'devil-kani'. If a water-saturated layer of soil lies above a tunnel, it is obvious that we cannot dig a well through it, because the water would seep into the well and fill it up. 'Devil-kani' involves digging the well from the tunnel to the surface so that water pours into the tunnel and drains out.

10. What is the gradient of the gallery?

The average gradient is 0.8/1:1000 m, though the qanats in mountainous areas are steeper than the ones in the plains. The lower the gradient in the gallery, the sooner the gallery and the earth surface intersect.

11. What does 'tahsoo-bonsoo' mean?

'Tahsoo-bonsoo' is the method used is to correct the direction of the tunnel when the outlet of the qanat is different from the plan. Then we dig another tunnel from the planned outlet to the actual outlet, so that the new tunnel channels water according to plan.

12. How do you deal with blockages or dangerous gases in a tunnel?

Today we use electric drills but in the past we had to go through an exhausting process of digging through it, often going through several picks in the process and using two workers digging together. When it comes to gas, we used to hold a piece of canvas on to our mouth and nose. Another method is to dig two parallel shafts about a meter apart but linked at different points to improve air circulation. We use electric fans today but we used to use leather bellows linked by a hose to the surface. Another option was to drop sand, vinegar and lime into the well.

13. How do you waterproof the water transfer section of a qanat?

To prevent seepage if a qanat runs through porous soil we cover the tunnel floor with soft clay mud and trample it so firmly that water can no longer seep through it.

14. What is meant by the 'irrigation cycle'?

When qanat water eventually appears on the surface people take turns irrigating their farm lands with it. A rotation period depends on the number of people requiring water, the crop and the amount of water available. It may be eight days for one crop but some crops, like onions, consume more water and so farmers may agree to reduce the period to half. The units of measurement vary from region to region. Here we use a scale known as 'saboo' or 'tasht', which equals ten minutes of irrigation time. We measure it with using a copper bowl with a tiny hole in its bottom that floats on the surface. The bowl takes ten minutes to fill with water and that unit of time is called 'saboo' or 'tasht'.

15. Why are qanats drying up and how can you rehabilitate those that have been abandoned?

Apart from deep wells, the main reason is that qanats are being abandoned by their owners because of quarrels between them and farmers over the money needed to repair them. Therefore we need to find out how much a dried up qanat will cost to rehabilitate before we try to do so, and that means finding out why it dried up in the first place. We need to make sure that there is a good supply of groundwater, and we should watch out for the boundaries of other qanats.

16. How can you control the flow of water in qanats in the winter?

I don't think that it is right to block a qanat in winter to stop water being wasted as the build-up of water can cause the tunnel to collapse. Also, damming the surface streams can help replenish groundwater resources, because runoffs trapped behind the dam can seep into ground and recharge the aquifer. Artificial recharge dams are very useful, especially if they are built upstream from the qanat. The effectiveness of the dam depends on its size as well as the texture of soil. A dam should be 500 m from the qanat, and no more.

17. Can deepening ever make a qanat (especially a mountainous one) independent of precipitation?

Precipitation affects the discharge of a qanat whether or not the qanat is mountainous, though drought affects the qanat on the plain less than the mountainous one. To reduce the effect of drought, we extend the tunnel to drain a larger area of the water bearing zone, and construct dams to collect melt water and replenish the aquifer. Dredging is another procedure to save existing water, but extending can increase the discharge of a qanat.

18. Does it make economic sense to maintain and rehabilitate qanats?

It is better to rehabilitate a qanat than drill a deep well if the qanat is not beyond repair. A well requires fuel, maintenance of its motor and other inputs whereas a qanat flows as long as gravity permits. A deep well is like false tooth, but a qanat is your own tooth that matches your body.

19. How often do you have to remove the mineral sediment that builds up on the walls of the tunnel?

We have to remove the sediment in the water transport section but not in the water production section, because there is no build-up of sediment where groundwater seeps into the tunnel. So the presence of sediment can be used to determine the boundary of the water transport section. We strip the sediment from the walls of the tunnel just before it becomes so narrow so that nobody could pass through it.

20. How many people make up a qanat construction team?

Qanat construction cannot be done by a single person; at least two people are needed, more if you want to speed up the work. If the wells are far away from each other, we need five people, of whom two drag the debris along the tunnel, two dig the tunnel and the last person operates the well wheel to pull up the buckets of debris (Figure 6.18). If the qanat is deep, we add another person to help pull up the debris.

21. Given that qanat engineering is difficult and dangerous, how can we encourage people to stay in and even take up the profession?

One way is to make it public that our job is valuable and useful. Once the mullahs used to do this; they used to preach in the mosques that our job deserved respect not only in this world but also in the next. This encouraged people to take an interest in qanat-related work in spite of the danger. Another way to encourage qanat practitioners is to raise their wages.

22. What tools do you use?

We use something called a 'chapar' to protect us if the sides of a well start crumbling. It looks like an umbrella made of wood. Workers also wear a white hat padded with cotton. We use several types of shovel and pickaxe

Figure 6.18 How to operate the well wheel.

to dig the tunnel – a pickaxe for the tunnel, a heavy pickaxe for digging the well and an even heavier pick (that is four times heavier than the tunnel pick) for dredging the bottom of the tunnel. This pick has a long handle so that we can work in water, whereas the pick that we use to dig in hard rock has a short, thick point like a sparrow's beak. We use a well wheel to pull the debris from the well. This tool has been in use for centuries and consists of two vertical crosses with the arms linked by four horizontal poles and a rope coiled round it. There are two types of well wheel; the large one that is used in a water bearing zone, and the small one that is used to hoist dry debris. Four meters of rope coils once around the well wheel so we can estimate the depth of a well by counting the rounds of the well wheel. We use different types of rope, but the best is known as 'sazoo', which is made of palm fiber. This is strong, durable and does not rot in water. However, some people have started to use rope made of cotton, steel wire and plastic. We have a special measuring tool called a 'ragham', a wooden frame that we use to measure the discharge (Figure 6.19).

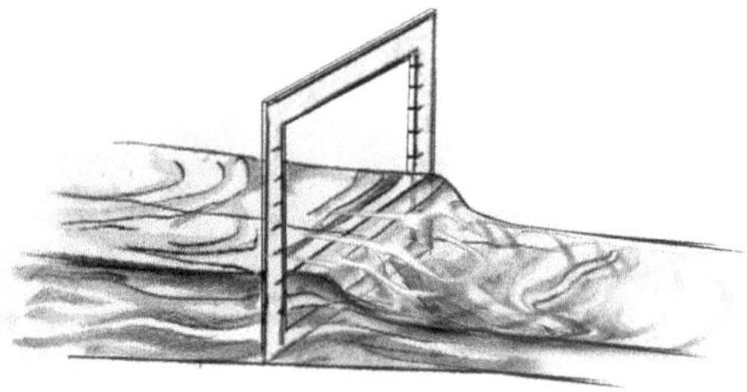

Figure 6.19 Wooden weir to measure the water discharge.

Figure 6.20 Determining the position of the stands of the windlass.

23. How do you reinforce the stands of the well wheel over the shaft?

To determine the position of the stands, somebody takes three small stones and drops one of them in the middle of the well (Figures 6.20 and 6.21). He spreads his arms and drops the other two stones at the same time and uses that to determine the position of the well wheel. We gouge two holes in both sides of the mouth of shaft across from each other and put the stands into the holes and tamp them with stones and dirt.

24. How do you light the tunnel?

We used to use a small bowl full of vegetable oil in which there was a wick. But this was replaced by the carbide lamp. We used to hang the lamp from the wall at shoulder level to the left and right while digging a tunnel. Digging a well does not need any light, and a skillful worker can dig even in the dark.

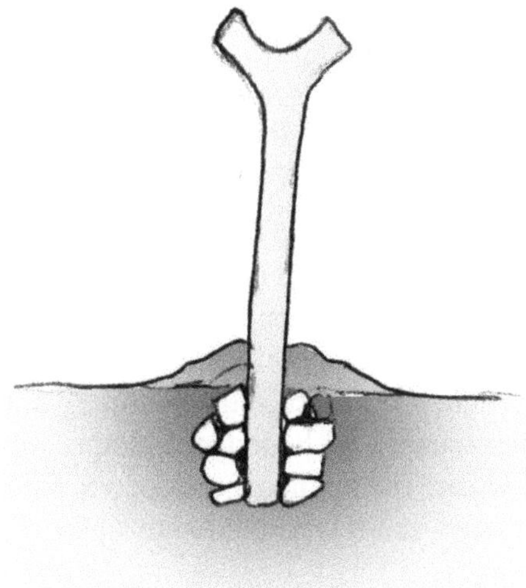

Figure 6.21 Fixing the stands of the windlass.

25. What do a 'mirab' and an 'abyar' do?

A 'mirab' is responsible for distributing water. He should be a wise Muslim and known and trusted by everybody; he can be dismissed by the qanat owners if he makes a mistake. The 'abyar' channels the water from ditch to ditch to irrigate the farms. These two receive wages, but being 'mirab' is a privilege. The wages are not paid from the 'nafaghe', which is the money collected by owners to fund the repair and maintenance of the qanat, but they receive tips and gifts.

26. What do you do with owners who refuse to pay their share?

If somebody refuses to pay his share, he may be deprived of his right of irrigation. But if he is bankrupt or in need, somebody else would pay his share for him. The 'maqsam' is a place where water is distributed. If the qanat does not have enough water, we collect its discharge in a pool before distributing it to the shareholders. If there is more than enough and can be distributed among several shareholders at the same time, we build a 'maqsam' across the flow.

27. How do you finance the maintenance of qanat?

In the past the landlord would finance the construction of a qanat and the farmers would rent the water. But in many cases owners overcharged the farmers and the additional sum went to the maintenance of qanat. If a qanat needed to be extended, the renters had to contribute additional funds. The amount of money each farmer had to pay was in proportion to the shares of water he rented.

28. How do you determine the limits of a qanat when another qanat is to be constructed nearby?

The boundaries of a qanat depend on the characteristics of the soil through which the qanat is to be dug and the depth of the shaft. The length is between four and 8.5 m, a distance known as 'kolang andaz', which means 'throwing the pickaxe'. In the past a qanat engineer would stand by the well and throw the pickaxe and the spot where it fell marked the boundary of the qanat. When a qanat is dug below other qanats, the boundary is larger otherwise the new qanat would drain all the water. The boundary area of a qanat running through cultivated lands is three meters and it is forbidden to sink a deep well upslope from the qanat.

29. Should we use abandoned qanats to dispose of urban wastewater?

No! Never! People are entitled to their qanats, and no one has the right to appropriate qanats for public profit without the permission of the owners.

Fallah and Nejati

Mr.Nejati who comes from the township of Abarkuh was born in 1941 & has always worked with his old friend Mr.Asad Gholi Fallah. They have inherited this profession from their fathers.

1. How do you locate a water bearing zone in a desert?

There are many signs – for example the type of plants and soil, and signs of a river bed – that indicate the presence of groundwater. When we have found the spot we dig a borehole, the 'gamaneh' and then we measure the water seeping into the well, as well as the quality of the water itself. If this is adequate, we sink the other wells and the tunnel.

2. Does the water transport section differ in length according to the season, and is it possible to measure discharge because of differences in seepage?

The water transport section lengthens during a drought. As for measuring the discharge, we know that we lose about 25 percent of the discharge along the tunnel, although this varies slightly depending on the type of soil and the age of the qanat. The lower the water loss, the older the qanat.

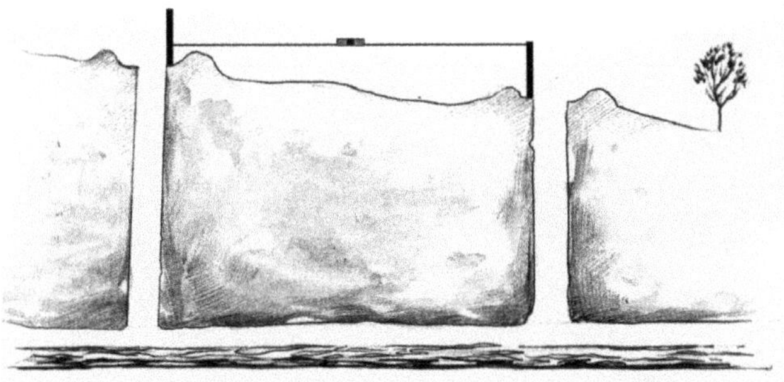

Figure 6.22 How to determine the depth of the next wells by using leveling tool.

3. How do you determine the length of a qanat and locate its outlet?

First, we calculate the difference between the levels of water in two wells, and then we work out with a leveling tool (Figure 6.23) by how much the spot where the borehole has been dug is higher than the spot where the gallery is to exit. This is how we know the exact location of the outlet of the qanat (Figure 6.22). The shafts of the qanat vary between 60 to 100 meters; although if there are poisonous gases in the gallery, we have to dig the shafts nearer to each other.

4. What is a 'dastak' and what is a 'harang'?

'Dastak' refers to the two side branches leading from the two sides of a main tunnel to bring more water to the gallery. When a large rock blocks the tunnel, we bypass it by digging above it (Figure 6.24) and the waterfall that results from the water reaching this point is called a 'harang'.

5. What is the diameter of the shaft and what are the dimensions of the tunnel?

Round wells are 80 cm in diameter and rectangular ones are 60 × 120 cm. It is easier to work in a rectangular well than a round one, because you can go up and down on one side of the well, while the bucket of debris goes up and down on the other. But round wells are more resistant to cave-ins.

The tunnel is usually 1.5 m high and 60 cm wide, to allow men and water to pass along it, with an average gradient of 20:1000, or enough to maintain constant water depth. The roof forms a crescent to channel water down the sides of the tunnel. If we are digging a tunnel through water bearing rock, we use a tool called a 'rassi' to find the point where we begin to dig up to the surface, an action called 'devil'.

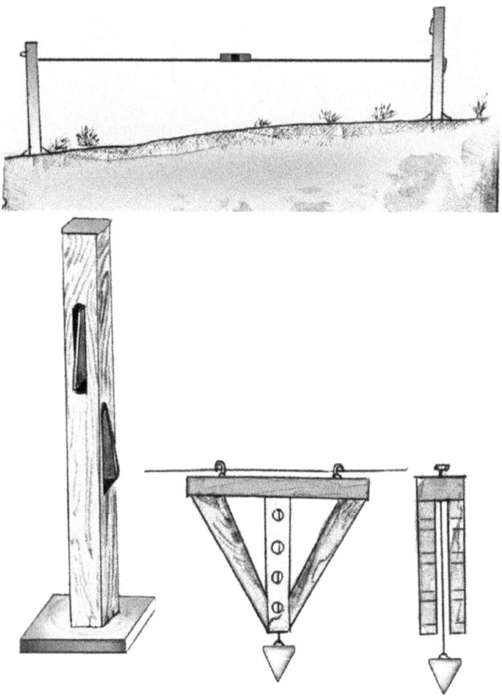

Figure 6.23 Close-up picture of a traditional leveling tool.

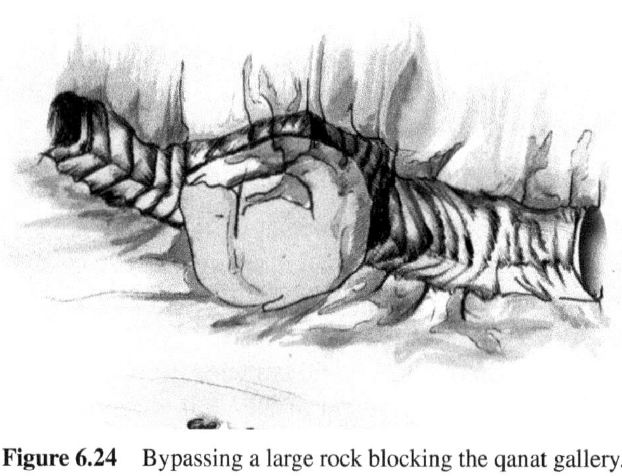

Figure 6.24 Bypassing a large rock blocking the qanat gallery.

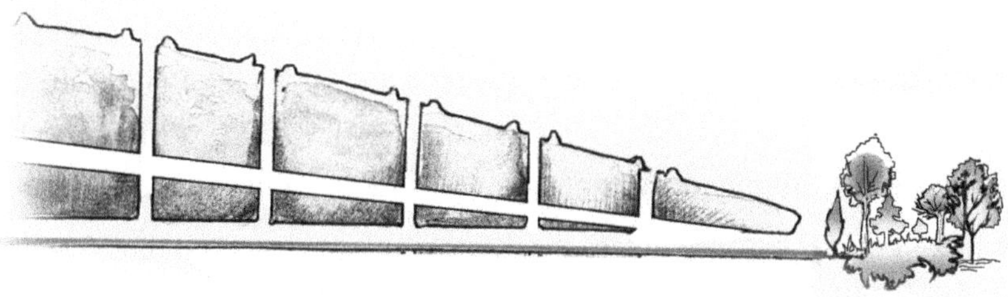

Figure 6.25 Deepening a qanat gallery by 'tahsoo-roosoo' method.

6. What does 'tahsoo-roosoo' mean?

When an aquifer is depleted we have to lower the floor the gallery to the new level of the water table. If we have to dig more than 1.5 m, we abandon that tunnel and dig a parallel tunnel under it. The old tunnel is called a 'roosoo', and the new one is called 'tahsoo' (Figure 6.25).

7. What do you do if you notice harmful gas in the qanat?

Nowadays we used electric fans to ventilate the tunnel and the well, but in the past we used leather bellows attached to a hose leading to the surface. Another way was to drop a mixture of sand, vinegar and lime into the well to displace the gases.

8. How do you prevent surface runoff from entering the qanat?

We line the gallery with ceramic rings called 'kavals' as well as limestone, and build a protective wall up to two meters high around the shaft and also cover it and cover around and over the mouth of the shaft.

9. How long is the irrigation cycle?

In Abarkooh the irrigation cycle lasts between 12–15 days; in most other regions it was 12 days but qanat owners have extended this to 15 days to maximize their profit. There are different words to define the units of time that divide the irrigation cycle. In Abarkooh, the word 'tashte' is used to mean a 10 minute period of irrigation, and the word 'habbeh' is used for an hour of irrigation time.

10. Can we store winter rainfall in qanats?

I don't think that damming the water is good for qanats because of the build-up of pressure on the walls. It is better to allow the water to drain out all the time – it is never wasted because even in the winter we have lands that need irrigation. We need to preserve our qanats because they are better for the environment and more sustainable than pumping wells. They are expensive to construct, but the benefits they bring to the land and rural communities more than offset the cost.

11. Which is more important, dredging a tunnel or extending it?

Dredging of course, because unless the tunnel is clear, water will not flow no matter how much extension work is done. When a tunnel is filled with debris we use a frame to reinforce the sides and ceiling and then pull the debris out. Mineral deposits, called 'shaa', also build up on the sides of the tunnel and have to be removed periodically.

12. How big is the qanat construction team?

Normally four people make up a qanat construction team, of whom two work on the surface and the others underground. But we will add more if we need to.

13. What tools and equipment do you use?

We use a heavy pickaxe weighing about 10 kg, which is called abgirkani and a lighter one weighing 4 kg called a 'kolang-e pishkarkani'. Other kinds of pick are also used – these include the 'bargbidi' and 'soozani'. We also use a well wheel called a 'charkh-e chah' that consists of two vertical crosses whose arms are linked by four horizontal poles with a rope coiled round it attached to a bucket – or to one of the workers. We fix the well wheel to the ground by digging two holes 30 cm deep on both sides of the well and putting the stands of the well wheel in these and tamping them with stones and dirt (Figure 6.26). We rotate this wheel to bring debris or people to the surface. There are two types; a large one used in a water-bearing zone, and a small one used to hoist the dry debris. The bucket, called a 'dalv', is made of leather; we hem the mouth with a round stick so that it is kept open and day we soak it in a lime solution to stop it shrinking. Traditionally we used rope made of palm fiber called 'sazoo', which is strong but needs to be kept wet, but nowadays we use cotton, which is more expensive and not as strong.

14. How do you light the tunnel?

In the past the only way to illuminate the tunnel was to use a small bowl full of vegetable oil in which there was a wick. But now we use a carbide lamp. We hang the lamps alternately on the left and right walls of the tunnel at shoulder level. Of course we don't need lamps when we dig the well.

Figure 6.26 Well wheel or windlass for hauling debris on to the surface.

15. What does the 'mirab' or 'sartaq' do?

In Abarkooh we don't have these positions; here the 'dashtban' is in charge of the distribution of water among the farmers. He is someone who is trusted by everyone to distribute the water fairly. He used to be paid at harvest time; he would go to the central threshing area and spread a cloth on the ground so that people could give him a portion of their harvest. Nowadays he is paid in cash.

16. What does 'nafaqe' mean?

The 'nafaqe' was a special fund the landlords spent on qanats. Sometimes it was a form of tithe, portions of the harvest of qanat users that was stored in an 'anbar-e nafaqe'. But now the word 'nafaqe' refers to a sum of money collected from all the shareholders of a qanat to repair, extend and maintain the qanat. If someone refuses to or cannot pay his share, he can borrow money from another shareholder or sell his share.

17. What do you do if you have to cross someone's farm while digging a qanat?

We try to get permission but if a landowner tries to stop us with no reason, we carry on regardless – the law is on our side.

18. What is the protected vicinity of a well?

This is the area around the well that cannot be dug or used. It is known as 'kolang andaz' and it is an area of around five meters, or the distance that someone could throw a pickaxe, around the well.

19. Can you run waste water pipes along dry qanats?

Yes, in fact it is a good idea use of dry qanats to carry sewage as long as a sealed pipe is used to protect the groundwater. We also use dry qanats to get rid of urban runoff as long as it does not pollute the aquifer.

Seyyed Jalal Hashemi

This veteran practitioner who has born in 1944 lives in Rezvanshahr has worked in Qanats since 50 years ago. His ancestors have all been Qanat practitioners. He has worked in Golestan & Lorestan provinces for a long time.

1. How do you know where is suitable for digging a qanat?

All we need is to know if where we are going to dig the qanat has an appropriate gradient or not.

2. After you figure out where to construct the qanat, what is the next step?

The next step is to sink a trial well to see how deep the groundwater is. We can predict where the outlet would be by knowing the depth of aquifer. It would be up to a skillful qanat master to locate a trial well which should be at least 1500–3000 meters away from the nearby qanats.

3. How do you know which part of a tunnel is the water transport section?

Wherever lacks the seepage is called water transport section.

4. How do you rate the discharge of a qanat by sinking a trial well?

To make sure that a qanat would drain a good supply of groundwater, we manage to sink another trial well just 1000 meters upslope from the first one. If we content ourselves with a trial well, it would be likely the well would hit a small confined supply of water misleading us. To prevent such a mistake, we prefer to sink another trial well to see if there is a reliable groundwater resource being worth spending too much time and labor. In case the amount of seepage in the two trial wells is the same, we can continue with the work, otherwise we had better give it up and find another site.

I should say that the trial well is sunk 0.5 meter below the water table to see how long it would take the seeping water to fill up this part of the well. Doing so, we can predict the output of the qanat.

5. How do you calculate the length of a qanat?

I want to explain this by giving an example. Imagine we have sunk a 30 meter deep trial well, and we are to dig the other wells every 50 meters until the outlet. To keep the tunnel horizontal, we measure the surface gradient, so that the bottoms of all the wells would be on a line parallel to horizon. Therefore, we may consider the depth of the next well 29 meters and so on.

6. What does the distance between the wells correlate with?

The distance between the wells correlates with their depth. In most cases the distance between two particular wells is twice their depth. In the past, if we wanted to dig through a ground giving off harmful gases, we had to sink the wells much closer together in order to better ventilate the tunnel. But at the present time, the modern devices have made it possible to consider the wells farther, even with the presence of the harmful gases.

7. How much should a well be in diameter?

A well is usually 80 centimeters in diameter, which may be decreased to 70 centimeters in order to speed up the work.

8. How high should a gallery be?

The height of the gallery varies from 1 to 1.5 meters. The minimum height is 1 meter, so that the worker has enough room to work in a sitting position. The water production section is constructed higher than the water transport section.

9. How do you know when a qanat should be cleaned?

That depends on whether the qanat is prone to collapse or not. The softer the ground through which the gallery cuts, the shorter the intervals between the cleanings. In case a flash flood can rush into a qanat, it would be inevitable to clean the qanat. In sum, the best way to know if it is necessary to clean a qanat is Ab-Harzi which means checking the gallery to see if there is any obstruction or not. Ab-Harzi can be done by four workers out of which two workers climb down a well and walk along tunnel while two other workers are keeping pace with them on the surface. When the two workers walking on the surface reach the next well, they send a rope down the well to haul the workers to the surface, and they swap over. While walking down the tunnel, the workers scan all over the conduit to spot any likely blockages or uneven places.

10. Could you explain what an irrigational cycle is?

An irrigational cycle is a rotation period during which the shareholders are supposed to take turns irrigating their lands with the water of their qanat. For example, if the irrigational cycle of a particular qanat is 16 days, it means that each shareholder has the right to appropriate the water just once every 16 days. The duration of the irrigational cycle varies from area to area with the geological conditions as well as the cropping pattern. As an instance, in case of porous soils into which water can seep quickly, the irrigational cycle tends to be shorter, so the plants can better cope with the shortage of water. However, a heavy clay ground makes the cycle longer, because such a soil can keep the water long enough for the crop to survive until the next irrigation.

11. Given the present situation, do you think it makes economic sense to build a new qanat?

Yes, it is still possible to go around building qanats, though the cost of this job is much higher than what done in the past.

12. Are the existing qanats worth preserving at the present time?

Yes, they are, for several reasons. One, all the qanats have long constructed and deserve to be treated as cultural heritage. Two, all the qanats can drain out groundwater just by the force of gravity with no need for any fuel or energy. Three, the water of qanat can be purified while flowing down the tunnel and coming in contact with soil.

13. We know that most of the qanats are subject to obstruction due to the build-up of the sediments, how often do you remove his sediment from the tunnel?

That depends on water quality and topographical condition. The qanats, running in mountainous areas leave more sediments than those running in plains, because the faster the water flows, the more sediment would be left.

14. What is a Maqsam like?

If two or more villages or agricultural areas are entitled to a particular qanat, a special structure named maqsam is built across the canal immediately after the water reaches the surface. In fact a Maqsam is a small dam with two or several outlets either of which directs water to a particular area. By means of a maqsam, it is possible to distribute water among the areas.

15. We know there is a custom of collecting money for qanat. What portion of this money comes from the owner and what portion comes from the renter?

There are 5 or 6 experts who gather once a year to decide the amount of money everybody should contribute. Those who use the water are obligated to take part in covering the expenses of the qanat, and Mirab is in charge of collecting this sum. If a qanat is landlord owned, a part of the money is up to the landlord to pay, and the rest should be paid by the peasants, otherwise the peasants are committed to pay the whole money on their own.

Abbas Nasiri

Haj Abbas Nasiri was 73 in 2001. His forefathers have all been practitioners for five generations. Mr. Nasiri has apprenticed this career to his father and uncle some 62 years ago. He has worked in Qanats in Yazd, Isfahan, Fars, Kerman, Hormozgan, Khorasan and Azerbaijan too.

1. How do you know where is more suitable for the building of a qanat?

We can find the best place from the condition of soil as well as the shape of the surrounding mountains. For example the sandy and lime soils can better keep groundwater. Also the discharge of the surrounding qanats can tell us if there is a good supply of groundwater here or not.

Altogether the master workers are quite familiar with the region in terms of having groundwater. For example everybody knows that the further you get away from the plain of Aqda toward Isfahan, the less chance you have to find a good supply of groundwater. Therefore locating a qanat depends on the master's ability to make use of his experiences in finding the best spot.

2. What do you do after you achieve the best place?

We start to dig a trial well 80 centimeters in diameter to find out how deep the water bearing zone is. Taking into account where this water is to be directed, if the trail well is too deep, you have to give it up, because if you get started on the horizontal tunnel from the bottom of such a deep well, the groundwater will not turn up at the right place.

3. How do you know if a qanat will have a satisfactory discharge or not from its trial well?

As I told you, the trial well should reach the groundwater at an appropriate depth. After we reach the groundwater, we keep going down until the seepage is too much that we cannot continue. Now an experienced worker can estimate the output of this qanat from the amount of seepage. If you have to haul a bucket of water along with 10 buckets of debris, it means that the trial well is short of water and the work is not promising. But if you haul 10 buckets of water along with a bucket of debris, your qanat will have a good discharge.

When we estimate the discharge of a qanat from its trial well, we take into account the length of the tunnel along which the water will flow, because it is inevitable to have some loss along the water transport section. All we can do to reduce the waste of water is to apply a kind of soft clay to the floor of tunnel in order to seal the tiny cracks through which water may escape.

4. How do know how long the qanat would be, after finishing its trial well?

It is easy to estimate the length of a qanat and its outlet from the depth of its trail well and the surface slope. For example if the trial well is 50 meters deep and the surface gradient is 20 cm per 100 m, then the qanat would be 25 km long. To more accurately estimate the surface gradient, we use a very simple tool consisting of two 1.2 meter tall poles with a rope tied to the top of both. We hold the two poles away from each other in vertical position, and then we move down one end of the rope until the rope is quite horizontal. Imagine we have a 50 m long rope whose one end has been moved down just 10 cm, so the surface gradient would be 1/500.

5. How far the shafts should be from each other?

That depends on the type of soil in which the shafts are sunk as well as the depth of qanat. In general the distance between the shafts should be twice as much as their depth. Note that if the soil gives off too much gas, we have to envisage some additional shafts to facilitate ventilation.

6. How do you correct the direction of the tunnel in case of any deviation?

If a tunnel deviates from the expected direction, there are two ways to put it right:

A) Dastak Zadan – B) Kaje Kardan (Figure 6.28)

Dastak Zadan is a term for the procedure of bypassing a missed shaft. The picture below shows a tunnel deviating from the way to the shaft number 2 (Figure 6.27). In this case we keep on digging up the tunnel in a round way to the shaft 3, but we connect this lateral tunnel to the shaft 2 in order to make use of the shaft 2 through which the debris can be hauled to the surface.

Kaje Kardan refers to the procedure of correcting a wrong direction. The following picture shows a deviation from the expected direction, but the wrong tunnel has been put right by being tilted toward the shaft 2.

7. How high should the gallery be?

The height of the gallery depends on the method of digging and the type of ground. If the gallery is dug by compressor drill, the gallery should be high enough so that the machine has enough room to work. In this case the gallery should be at least 2 meters high. Also if the shafts are so far away from each other, the gallery should be relatively high to offset the lack of ventilation.

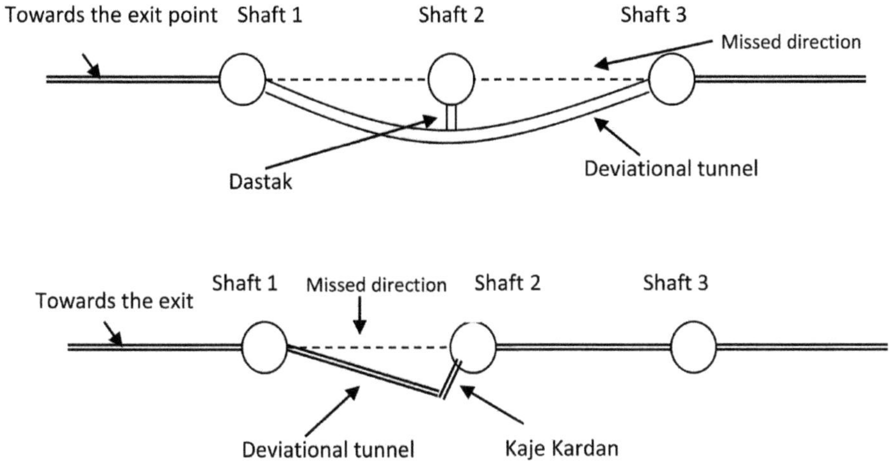

Figure 6.27 Methods to dig a deviational tunnel.

Figure 6.28 Dastak Zadan and Kaje Kardan: two kinds of abnormalities in the direction of gallery.

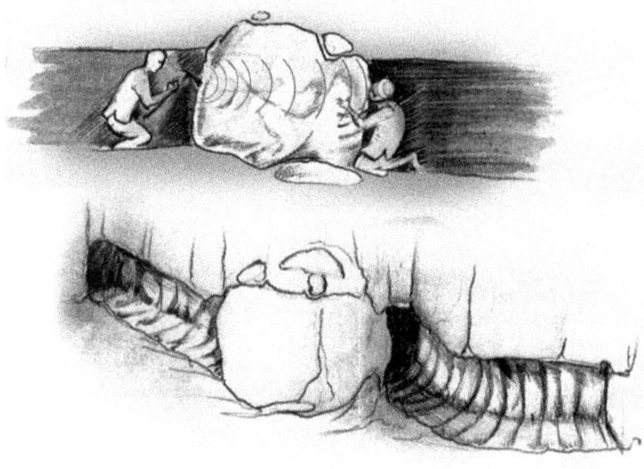

Figure 6.29 Digging through a rock.

Figure 6.30 Devil zani; method of digging overhead.

If we have to dig through a huge rock, we do not dig the gallery higher than 80 centimeters (Figure 6.29).

8. In case a tunnel runs below a saturated layer, how you can get the shaft to such a tunnel?

Just by digging a Devil that is the method of digging a well from down to up. If a water-saturated layer of soil lies above a tunnel, it is obvious that we cannot dig a well through it, because the water would seep into the well and fill it up. Devil involves digging the well from the tunnel to the surface so that water pours into the tunnel and drains out (Figure 6.30). Bizesh is another type of Devil, dug diagonally where the soil is so hard that the worker cannot dig straight overhead (Figure 6.31).

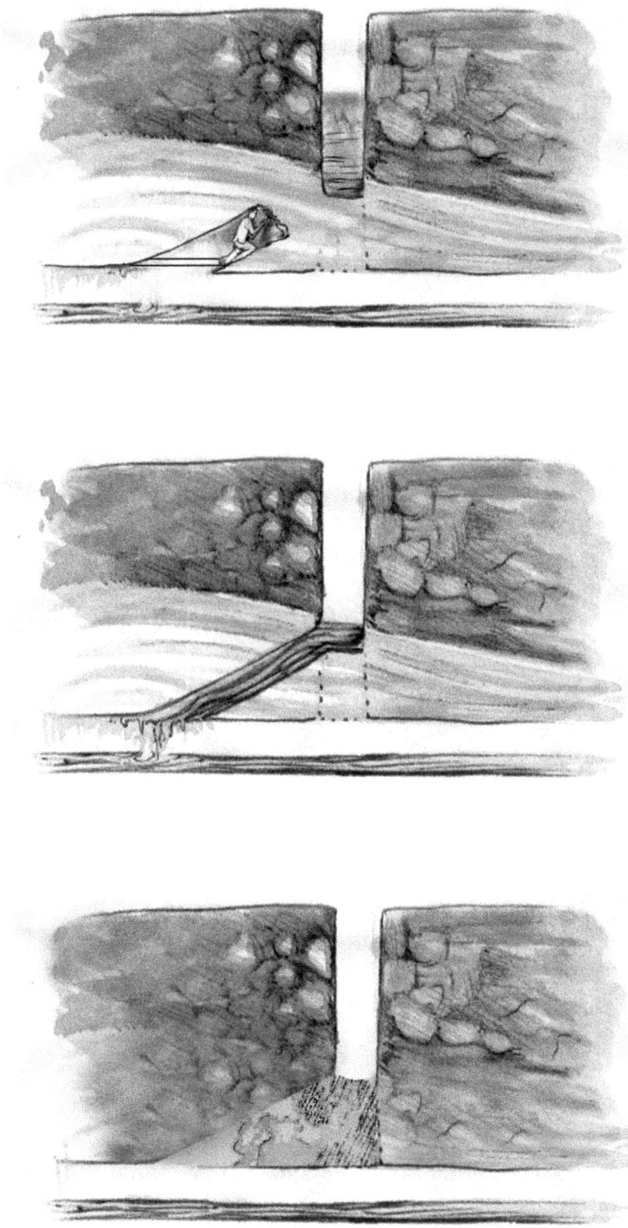

Figure 6.31 Bizesh: A kind of devil which is dug in the hard soil.

Figure 6.32 U-shaped tunnel in which water flows in one end and forces water up and out the other end.

9. Is there any method to guess the quality of water before digging a trial well?

It is obvious that in a brackish ground our qanat will bring salty water which may be of use to some crops such as barley, but bitter water is completely useless.

10. If there would be another qanat on your way, how do you pass it?

If the two qanats are at the same level, we have no way but building an inverted siphon called Shotor Galoo. At this point we dig a U-shaped tunnel to avoid intersecting another tunnel (Figure 6.32). Water flowing in one end of this siphon simply forces water up and out the other end. This is especially important in qanat systems which must be routed under rivers, other qanats or other deep obstructions.

But in case two qanats pass at two different levels and through a soft crumbling ground, all we should do is to seal the ceiling of the lower qanat with brick, stone and cement.

11. Could you explain the advantages of a trial well?

The first advantage is to find out about the quantity and quality of the future qanat. The second is to specify the vicinity of the qanat to make sure that nobody else would go around building another qanat in this vicinity. Besides if the workers need any water for drinking, they can use this well.

12. What remedy do you have for the problem of harmful gases in qanat?

There may be two kind of gas in a qanat: the first one is a gas that suddenly rushes into the tunnel and overruns everywhere immediately. The best sign of being such a gas is a light going out. As soon as we notice our light is going out, we get out of the qanat. This kind of gas is temporary and vanishes after a while. To find out if the gas is gone or not, we tie a burning light to a rope and send it down the well, if the light is out it means that we should still wait for the gas to disappear.

Another kind of gas is that we call Gaz-e Damz. When we dig through the ground, we may come across an area that constantly emits this gas that does not put out our light especially in small quantity. When we find it hard to breathe we notice that there is such a gas in the tunnel. There are some ways to get rid of this gas, for example pouring some damp sand down the shaft. This practice makes the air in the tunnel circulate. To do so, we can burn a ball of dry thorns and drop it down the well. We may pour some vinegar into the shaft as well.

13. How do you prevent the surface floods from rushing into the shafts?

The first way is to block the shafts by means of clay, lime and brick at a spot 2–3 meters below the earth surface. As you know, when we haul the excavated materials to the surface, we dump them around the mouth of well to build a barrier against the flash floods. This barrier works to protect the shafts from the seasonal runoffs, unless the fury of water is so much that we have to seek a better solution. In this case, we had better build a wall with cement and brick around the mouth of the shaft (Figure 6.33). This wall should be so strong that it does not give way easily under the fury of water.

14. When do you decide to clean a qanat?

Every qanat needs to be dredged once in a while, because the tunnel may crumble and As a result, some dirt and silt may accumulate on the floor and gradually obstruct the qanat. The sediments may worsen the situation as well. It depends on the condition of soil and the amount of discharge to determine how often a particular qanat requires to be cleaned. Besides, when we are in the process of extending a qanat, it is necessary to clean the downstream, for the water is do muddy that some deposits are inevitable. Also, if a flood can penetrate the qanat somehow, we have to fix all the destructions it has brought about.

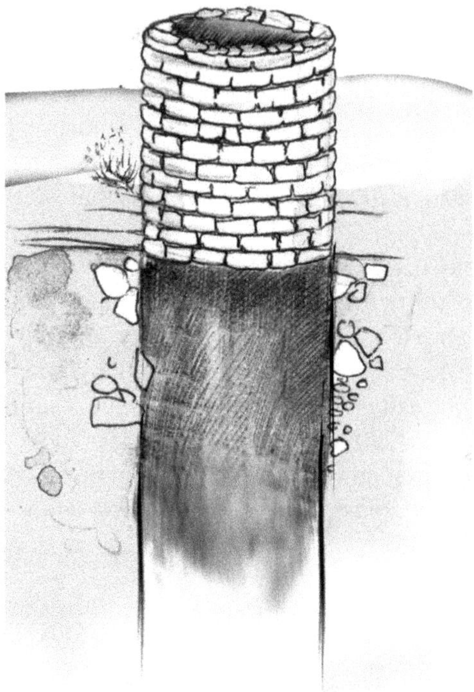

Figure 6.33 Wall with cement and brick around the mouth of shaft to protect it from the seasonal runoffs.

15. Could you explain what an irrigation cycle is?

An irrigation cycle is a rotation of irrigational rights during which the farmers take turns directing the water of a qanat to their farm lands. For example if an irrigation cycle is 8 days, you have the right to use the water of qanat just once every 8 days. The length of an irrigation cycle is dependent on the prevailing cropping pattern in the region. For example if the most parts of the area has been devoted to the cultivating of vegetables or fruit trees, the irrigation cycle would be 8 days, because the longer intervals may damage such plants which should be watered more frequently.

16. Based on what scale you divide water among the shareholders?

In terms of the pumped wells, we use Hour to calculate the time of irrigation, but in terms of qanats it is common to use the term Jorreh as a unit of time. Each Jorreh equals 11 minutes calculated by a traditional water clock (clepsydra) consisting of two bowls. The smaller bowl has a tiny hole at its bottom. When this bowl is placed on the water surface, water starts to enter the bowl through its hole until the bowl is filled up. It takes 11 minutes (or a Jorreh) for the bowl to sink in water. Note that there is a person named as Tayyare Gardan who is in charge of checking up on the small bowl to see if any residue has been lest on the rim of the tiny hole. He runs a thread into the hole of the bowl and moves it forth and backward gently in order to clean the rim of the hole, because even a little sediment can decrease the water coming in and it would be to some of the shareholders' detriment.

17. In your opinion, what are the main causes for the annihilation of qanats?

The first cause is the lack of cooperation and coordination between the shareholders. Some of the shareholders refuse to contribute to collect the money needed to repair or maintain their qanat, so those who may contribute would lose their motivation to keep spending money on the matter from which others would profit with no pain. Therefore the qanat will be left deserted. The second cause that deserves to be mentioned is the pumped wells that have mushroomed in the recent decades. Many of these wells have been drilled in the bound of the qanats which never compete with the electric or diesel pumps. A drought may aggravate the problem as well, where an aquifer is short of recharge.

18. Could you explain what a Goorab is?

Goorab is an earth dam or dike to collect the seasonal runoff to recharge aquifers. In mountainous areas, we build the dam about 150–200 meters upstream from the deepest and last well of the qanat so that the water trapped behind the dam will seep into the earth and increase the discharge. The dam must be far enough away from the qanat so that it does not affect its structural integrity. Also, damming the surface streams can help replenish groundwater resources, because runoffs trapped behind the dam can seep into ground and recharge the aquifer. Artificial recharge dams are very useful, especially if they are built upstream from the qanat.

19. Does the deepening of qanat work to reduce the fluctuation of its discharge from year to year?

I don't think so. If you want to deepen a qanat, you have to do this business along its tunnel all the way. Who can repeat such an exhausting labor every year keeping pace with the groundwater table constantly going down? So to keep the discharge steady, you had better extend the tunnel horizontally through the water bearing zone.

20. What we can do to rehabilitate the qanats that have dried up because of the pumped wells drilled in their vicinity in the recent decades?

All you can do is to extend these qanats to reach a new supply of groundwater as long as there would not be any other pumped wells on your way.

21. We all know that you are involved in a very dangerous job, how one can encourage your colleagues not to give up this profession?

This profession has been passed from father to son, but now our sons no longer want to keep up their father's job, for they have no insurance despite lots of dangers they have to deal with.

22. Taking the present situation into account, does it make economic sense to keep maintaining the qanats?

Sure! For this system provides a constant energy-free discharge as well as some job opportunities. Also, the water coming through a qanat is more useful to the crops than the water being extracted through a pumped well. My own experience shows that in case a farm is irrigated by the water of a qanat produces more crops than when irrigated by the water of a pumped well. I think it is attributable to the minerals and microelements that water brings along running through many types of soil in a qanat.

23. Is it right to use the existing qanats to replenish the aquifer?

Just in terms of the dry qanats, you can direct the flood to the shafts.

24. How often should a qanat be cleaned to remove the sediments from on its floor and sides?

That depends on the type of soil the qanat cuts through and the minerals dissolved in water, so you may find a qanat that does need to be cleaned sooner than once every ten years. Actually, the faster a flow, the more the water would leave sediments in the tunnel.

25. How do you measure the volume of a flow?

To do so, we have a scaled frame. First of all, we level the bottom of the ditch just where we are to measure the flow, because to get an accurate result the slope should be zero and water moves very slowly through the scaled frame. Now with the help of the marks engraved on the frame we can measure how much water if flowing (Figure 6.34).

Figure 6.34 Measuring water flow.

26. How do you designate the bound of a qanat?

The extent of the bound varies from qanat to qanat. It depends on the type of soil a particular qanat cuts through.

27. What if you have to dig across a farm land or a property?

We have no way but negotiating with the owner of the land to come to an understanding. We need about 5 square meters of his land for the digging of each shaft, and then we put a price on the area we need. We do our best to agree on the price and settle the problem. If we failed to persuade the owner, we have to request a court to intervene.

28. How far from a qanat it is possible to drill a pumped well?

It is okay to drill a pumped well somewhere downstream from the outlet of the qanat, for this spot has nothing to do with the water supply being drained by the qanat. No doubt the drilling of such a well upslope from the mother well of the qanat can inflict a great damage on the discharge of the qanat, because the pump sucks out the groundwater even before it reaches the gallery of the qanat.

29. Is it right to use the abandoned qanats to convey sewage?

If we can seal the tunnel so firmly that no sewage can leak out, it's okay, but in my opinion we had better let go of this idea, because in any way there is a risk of polluting the groundwater.

References

Dart, R. A. and P. Beaumont, (1971). "On a Further Radiocarbon Date for Ancient Mining in Southern Africa" South African Journal of Science, January 1971, pp. 10–11.

Morton G. R. (1996). Subterranean Mining and Religion in Ancient Man, http://home.entouch.net/dmd/mining.htm

Semsar Yazdi Ali Asghar, (2004). 'Qanat from practitioners' point of view (Farsi version), Iran water resources management company.

Index

About the Authors

Ali Asghar Semsar Yazdi was born in 1956 in Yazd, Iran. He graduated from the Institute of Applied Sciences in Lyon, France (INSA de Lyon), with a PhD degree in civil engineering (1995). He has taken part in setting up such water organizations as Yazd Regional Water Authority, Yazd Water Museum, Qanat College, and International Center on Qanats and Historic Hydraulic Structures (UNESCO-ICQHS).

In 2000, he held the first International Conference on Qanat, and in 2003, he was awarded the "best researcher". From 2006 to 2013, he was the Director of the International Center on Qanats and Historic Hydraulic Structures. He is now senior advisor to this center. In February 2012, he organized the International Conference on Traditional Knowledge for water resources management (TKWRM2012).

He has authored and co-authored tens of papers and books on water management and Qanat system. Some of these books are as follows:

- Qanats of Bam from technical point of view (2005),
- Qanat from practitioners' point of view (2010),
- Qanat in its Cradle (2012),
- Qanat of Zarch (2014),
- Qanat Tourism (2015),
- Qanats of Emamieh and Qasemabad (2016), and
- Qanat Knowledge: Construction and Maintenance-Springer (2017).

Majid Labbaf Kheneiki was born in 1975 in Mashhad, northeast of Iran. He received his bachelor degree in agricultural engineering, and then he continued his study in human geography at University of Tehran. He holds a PhD in human geography from University of Tehran. From 1998 to 2005, he worked as a researcher for the Amirkabir Research Institute and the Iranian Academic Center for Education, Culture & Research (ACECR). From 2005, he has been working as a senior expert and researcher for the International Center on Qanats and Historic Hydraulic Structures, a UNESCO Category 2 center. He has authored and co-authored the books Qanat Knowledge, Qanat Tourism, Qanats of Qasem Abad and Emamiyeh, Qanat in its Cradle, Water Division Systems in Iran, Water and Irrigation Techniques in Ancient Iran, Qanats of Taft, Qanat of Gonabad as a Myth, and Qanats of Bam from Technical and Engineering Point of View. Tens of his articles have appeared in Urban Tourism Journal (University of Tehran), Geographical Researches Journal, Mashhad University Journal, and so on.

For Product Safety Concerns and Information please contact our EU representative GPSR@taylorandfrancis.com
Taylor & Francis Verlag GmbH, Kaufingerstraße 24, 80331 München, Germany

www.ingramcontent.com/pod-product-compliance
Ingram Content Group UK Ltd.
Pitfield, Milton Keynes, MK11 3LW, UK
UKHW022210190926
475702UK00003B/290
* 9 7 8 8 7 7 0 2 2 0 8 4 2 *